Second Edition

Introduction to
Spices, Plantation Crops, Medicinal And Aromatic Plants

N. Kumar

B.Sc. (Hort.), M.Sc. (Ag.), Ph.D.,
Horticultural College & Research Institute
Tamil Nadu Agricultural University, Coimbatore

Oxford & IBH Publishing Co. Pvt. Ltd.
New Delhi
(*A Unit of* CBS Publishers & Distributors Pvt Ltd **)**

CBSPD

CBS Publishers & Distributors Pvt Ltd

New Delhi • Bengaluru • Chennai • Kochi • Kolkata • Lucknow • Mumbai
Hyderabad • Jharkhand • Nagpur • Patna • Pune • Uttarakhand

Introduction to Spices, Plantation Crops, Medicinal and Aromatic Plants

Second Edition

ISBN-13: 978-81-204-1776-2
ISBN-10: 81-204-1776-3

OXFORD & IBH
New Delhi
(A Unit of CBS Publishers & Distributors Pvt Ltd)

Published by **Satish Kumar Jain** and produced by **Varun Jain** for

CBS Publishers & Distributors Pvt Ltd

4819/XI Prahlad Street, 24 Ansari Road, Daryaganj, New Delhi 110 002, India.
Ph: 011-23289259, 23266861, 23266867
Fax: 011-23243014

Website: www.cbspd.com
e-mail: delhi@cbspd.com;
cbspubs@airtelmail.in.

Corporate Office: 204 FIE, Industrial Area, Patparganj, Delhi 110 092
Ph: 011-4934 4934
Fax: 011-4934 4935
e-mail: publishing@cbspd.com; publicity@cbspd.com

Branches

- **Bengaluru:** Seema House 2975, 17th Cross, KR Road, Banasankari 2nd Stage, Bengaluru 560 070, Karnataka, India
  Ph: +91-80-26771678/79   Fax: +91-80-26771680   e-mail: bangalore@cbspd.com
- **Chennai:** 7, Subbaraya Street, Shenoy Nagar, Chennai 600 030, Tamil Nadu, India
  Ph: +91-44-26680620, 26681266   Fax: +91-44-42032115   e-mail: chennai@cbspd.com
- **Kochi:** 42/1325, 1326, Power House Road, Opp KSEB, Power House, Ernakulum Kochi 682 018, Kerala, India
  Ph: +91-484-4059061-65,67   Fax: +91-484-4059065   e-mail: kochi@cbspd.com
- **Kolkata:** 147, Hind Ceramics Compound, 1st Floor, Nilgunj Road, Belghoria, Kolkata-700056, West Bengal, India
  Ph: +033-25633055, 033-25633056   e-mail: kolkata@cbspd.com
- **Lucknow:** Basement, Khushnuma Complex, 7 Meerabai Marg (Behind Jawahar Bhawan),Lucknow-226001, UP, India
  Ph: +0522-4000032   e-mail: tiwari.lucknow@cbspd.com
- **Mumbai:** PWD Shed, Gala no 25/26, Ramchandra Bhatt Marg, Next to JJ Hospital Gate no. 2, Opp. Union Bank of India, Noorbaug, Mumbai-400009, Maharashtra, India
  Ph: 022-66661880/89   e-mail: mumbai@cbspd.com

Representatives

- Hyderabad   0-9885175004
- Patna   0-9334159340
- Jharkhand   0-9811541605
- Pune   0-9623451994
- Nagpur   0-9421945513
- Uttarakhand   0-9716462459

Printed at Chaman Enterprises, Daryaganj, New Delhi, India

Dedicated to

Dr. C.R. MUTHU KRISHNAN, Ph.D.,
Dr. S. MUTHUSWAMI, Ph.D.,
Dr. K.G. SHANMUGAVELU, Ph.D.,

Former Deans
Horticultural college & Research Institute
Tamil Nadu Agricultural University, Coimbatore

Foreword

The vast area and the varied agro-climatic conditions of India ranging from tropical to temperate make it possible to grow almost all the different kinds of spices, plantation crops, medicinal and aromatic plants. India is the largest producer and exporter of spices, spice oils and oleoresins. Plantation crops, similarly, play an important role in our country's economy by earning valuable foreign exchange, providing direct and indirect employment to many people and also by supporting a number of horticultural based industries. Medicinal and aromatic plants in India are numerically very large, and provide the basic raw materials for the indigenous pharmaceutical, perfumery, flavour and cosmetic industries.

In view of the economic importance of these crops, research and development effort were focussed on these crops for the last three decades by the Government of India through the Central Research Institutes and Regional Research Laboratories and the State Agricultural Universities and as a result a significant stride has been made in respect of crop improvement, crop husbandry and crop protection in these crops. It is gratifying to note that the authors of this book have come forward to bring out a comprehensive book on these important commercial crops embodying these research results. The book contains four sections, each one dealing with a group of crops viz. Spices, Plantation Crops, Medicinal and Aromatic Plants with topics so arranged and constructed as to provide the complete details on their cultivation practices. It has been written in a simple language with adequate number of figures. This book will be very useful to the students of Agriculture, Horticulture and Forestry in the Agricultural Universities and Institutions where the curricula and syllabi include these crops and also as a hand book or reference book to the extension personnel of the State Agricultural, Horticultural and Forestry departments.

I congratulate the authors, Drs. N. Kumar, JBM. Md. Abdul Khader, P. Rangaswami and I. Irulappan of the Horticultural College and Research Institute, Tamil Nadu Agricultural University, Coimbatore for their effort to bring out this text book.

T.N.A.U, Coimbatore   **S. Sankaran**
10.8.93   Vice-Chancellor.

Preface to the Second Edition

The previous publication entitled "Introduction to Spices, Plantation Crops, Medicinal and Aromatic Plants" authored by me was well accepted and proved very useful to the students, teachers, extension specialists and also to the progressive growers of horticulture in our country. Subsequent to this first edition brought out in 1993, lot of new scientific information has become available in the last two decades due to the intensive research and development work carried out in these crops by various ICAR institutes and many State Agricultural Universities. This has prompted me to revise this present edition duly updating all the relevant information and also to include more crops.

I do sincerely hope that this new edition will again prove useful to the students, teachers, extension personnel of development departments besides to the enthusiastic growers dealing with this special group of horticultural crops. I am indebted to Dr. S. Marimuthu, Head (R&D), Parry Agro Industries, Valparai, Dr. Stephen D.Samuel, Entomologist, Regional Coffee Research Station, Thandigudi, besides to my colleagues in the University viz., Dr. M. Kalyana Sundaram, Prof (Entomology), Dr. N. Seenivasan, Asst. Prof (Nematology) and Dr. S. Harish, Asst. Prof (Plant Pathology) for their useful suggestion in the preparation of manuscript. I am also happy to record my appreciation for the help received from my PG Scholars particularly from Misses P. Janani, S. Aarthi, G. Amitha Kunikullaya during the revision of the book.

N. KUMAR
Coimbatore

19.02.2014

Preface to the First Edition

Horticulture deals not only with fruits, vegetables or flowers but also the other groups of crops namely spices, plantation crops, medicinal and aromatic plants. Spices and plantation crops play a key role in our economy as their export share to Agricultural sector is sizeable. Medicinal and aromatic plants are also economically important plants which provide basic raw materials for the indigenous pharmaceutical, perfumery, flavour and cosmetic industries. Our country is also earning precious foreign exchange from these group of plants.

In view of the economic importance of these crops, research and development efforts were focussed on these crops for the last three decades by various Central Research Institutes, National Research Centres, Regional Research Laboratories and State Agricultural Universities, Private Organisations like UPASI to alleviate many production problems. The valuable research findings emanated from these institutions have been disseminated through publication of research papers in various journals, research reports, proceedings of symposia and seminars. These informations are, unfortunately, sometimes beyond the reach of undergraduate students of Agricultural Universities, extension personnel in the development departments or enthusiastic growers.

In such a situation, the need for a simple book on these crops is highly felt by the students of Agriculture, Horticulture or Forestry as majority of

Agricultural Universities in India have these crops in their under graduate curriculum. This book is thus brought out mainly to meet the above requirement and is also the outcome of our teaching, research and field experiences in these group of crops for the last 2 to 3 decades.

The book has four major parts-Spices, Plantation Crops, Medicinal Plants and Aromatic plants dealt in order. Under each crop, with a brief introduction, aspects on botany, climate and soil, propagation, after cultivation, harvesting and processing including plant protections are dealt with. It is hoped that this book will be a boon to the students of Agricultural Universities, extension personnel in the State Departments of Agriculture/ Horticulture and also for the enthusiastic growers in our country.

We gratefully acknowledge the encouragement and enthusiastic cooperation extended by all our colleagues in the Faculty of Horticulture, more particularly by **Dr. R. Arumugam,** Dean, Horticultural College and Research Institute, Periakulam, **Dr. S.** Thamburaj, Professor of Olericulture, **Mr. R Venkatachalam,** Asst. Professor (Hort.), **Dr. V. Prakasam,** Asso. Prof. (Plant Pathology) and **Mr. P. Karuppuchamy,** Asst. Prof. (Ent.) besides by **Mr. P.** Hudson, Advisory Officer (UPASI) in the production this book. We are also grateful to Mr.S. Ravi Raj, Artist, TNAU, Coimbatore for bringing out the figures in a lively manner, sincere thanks are also due to Mr. C.S. Suresh for typing the original manuscript, M/s. Acme Computers, Madurai for the excellent typesetting of the manuscript and finally to M/s. Rajalakshmi Publications, Nagercoil for printing and publishing this book

We owe a debt of gratitude to **Dr. S. Sankaran,** the Vice- Chancellor, Tamil Nadu Agricultural University, Coimbatore for having adorned this book with his Foreword. We also acknowledge with thanks the permission accorded by our university in publishing book.

Sr. Author, Dr. N. Kumar wishes to record his special gratitude to his elder brother, Prof. N.T. Krishnan, to his colleague Prof Dr.G. Santhanakumar and also to Mr. S. Janakiraman, Tamilnadu Polytechnic, Madurai who are instrumental in bring out this book. The financial assistance extended **Dhanalakshmi Bank, Nagercoil** in printing this book is gratefully acknowledged.

We will be grateful to the readers if they help us with their valuable suggestions and construcutive criticism in the improvement of text for future editions.

Authors

Content

Part–I : Spices

Part–II : Plantation Crops

Part–III : Medicinal Plants

PART-I
SPICES

1

Introduction

Spices are those plants, the products of which are made use as food adjuncts to add aroma and flavour. **Condiments** are also spices, products of which are used as food adjuncts to add taste only. Both spices and condiments contain essential oils which provide the flavour and taste. They are of little nutritive value. They are used whole, ground, paste or liquid form, mainly for flavouring and seasoning food. Most spices increase the shelf-life of food, especially the dry varieties. Some are added to improve texture and some to introduce a palatable colour or odour.

There are about 63 spices grown in India and almost all spices can be grown in India because of the varied climate-tropical, subtropical and temperate prevailing in India (Fig. 1).

Classification of spices

Though spices can be classified in several ways based on

1. Plant part used-leaves, flowers, barks, rhizomes, fruits and seeds.
2. Botanical relationship - family to which it belongs.
3. Longevity of spices plants - annuals, biennials and perennials and
4. Morphology of aerial parts of spice plants - herbs with aerial stem, herbs with pseudostem, climbers, shrubs, trees etc.,

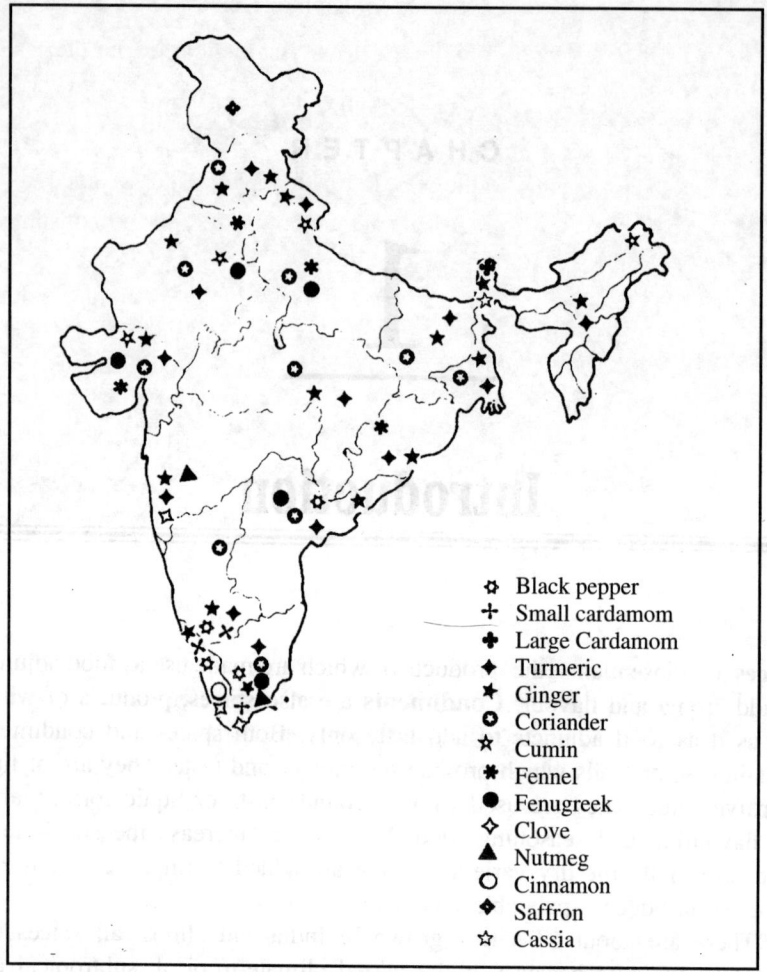

Figure 1 Places of cultivation of important spices in India.

None of the classification is complete as each classification has got some lacuna or over lapping.

Therefore at present a mixed combination of above classification is normally followed as below:

(a) **Major spices:** Cardamom, Black Pepper, Ginger and Turmeric

(b) **Seed spices** (Minor spices): Coriander, Fennel, Cumin, Fenugreek, Dill, Aniseed, Caraway, Celery and Bishop Weed.

(c) **Tree spices:** Clove, Nutmeg, Cinnamon, All spice, Cassia, Tamarind, Bay leaf, Curry leaf etc.

(d) **Herbal spices:** Rosemary, Thyme, Horse radish, Parsley etc.

(e) **Other spices:** Vanilla, Saffron, Asafoetida, Garlic etc.

Importance of spices industry in India.

1. No other country in the world grows as many kinds of spices as India.

2. India has been a traditional and largest producer, consumer and exporter of spices.

3. India produces spices on 3.08 million ha with an annual production of about 5.4 million tonnes valued at about Rs. 4500 crores, contributing nearly 20% of world's production, 30% of the trade in terms of quantity but only 10% in terms of value.

4. Nearly 90-95% of the total production is consumed locally and the rest exported.

5. India is the biggest exporter of spices and annually exporting about 4,45,000 tonnes of spices and spice products (Table 1.1) valued around Rs. 4,460 crores (contributing to about 12% of export earnings from agriculture or 1 % of total national export earning).

6. The export of spices and spice products is growing up every year.

7. They are indispensable part of our culinary preparations especially used for flavouring and seasoning of food.

8. In the daily intake of food and culinary preparations, the Indian food habits amalgamate divergent spices and exploit and utilize phytochemicals to add aroma and health in our daily life.

9. Most of the spices have potential medicinal values. Besides, the spices and spice products are also indirectly used as flavouring or colouring agents or as preservatives in many pharmaceutical preparations.

10. Spices have been used in cosmetic and perfumery industries. Spice oils are used in the manufacture of soaps, tooth pastes, talcum powder, after shave lotions, vanishing creams, mouth fresheners and room fresheners etc.,

Most of the spices are native of our country and hence India is aptly known as the **Land of Spices.** The history of Indian spices dates back to the beginning of human civilization and spices were mostly instrumental in the exchange of ancient culture and civilization within and outside the country.

Constraints in Indian spice industry:

1. Low farm productivity and consequent high cost of production

2. Non-coverage of high yielding varieties in farmers' fields
3. Non-availability of quality planting materials of high yielding varieties
4. Crop loss due to severe disease and pest incidence
5. Poor post-harvest handling
6. Inadequate extension network
7. Price instability
8. Stringent food laws by importing countries-Indiscriminate use of plant protection chemicals results in pesticide residues beyond tolerable limits (MRL) leading to rejection of many consignments of spices
9. Mycotoxins (Aflatoxin) contamination seriously affects the export potential of high-value commodity crops
10. Stiff competition from other producing countries -most of the exporting countries have no domestic market for the spices they are producing, forcing them to sell their produce even at cost price (cardamom from Guatemala, pepper from Vietnam, cloves & nutmeg from Indonesia).

Table 1.1 Export of spices from India

Item	Quantity (M.T)		Value (Rs.in lakhs)	
	1995–96	2010–11	1995–96	2010–11
Pepper	24,150	18,850	18,111.40	38,318.5
Cardamom(small)	500	1,175	1,239.55	13,216.25
Cardamom(large)	1,690	775	1,189.40	4,462.9
Chillies	55,200	2,40,000	19,084.00	1,53,554
Ginger	16,250	15,750	3,516.45	12,131.25
Turmeric	26,675	49,250	4,423.80	70,285.15
Coriander	10,700	40,500	2,075.20	16,663.25
Fenugreek	14,450	18,500	1,791.25	6,548.1
Other seed spices *	10,265	12,500	3,147.65	5558.05
Garlic	3,640	17,300	463.00	6,977.3
Other spices **	20,660	7,574	3,467.45	6,140.26
Curry powder	3,885	15,250	1,538.00	21,050.5
Spice oils and oleoresins	1,630	7,600	10,464.55	91,062.42
Total	1,89,695	4,45,024	70,511.70	4,45,967.93

* include cumin, celery, fennel, aniseed, ajwanseed, dill seed etc.
** include tamarind, asafoetida cinnamon, cassia, kokum, saffron etc.

2

Major Spices

2.1. BLACK PEPPER (*Piper nigrum L.*)

Family: **Piperaceae**

Black pepper, the **king of spices,** is obtained from the perennial climbing vine, *Piper nigrum* which is indigenous to the tropical forests of Western Ghats of South India. It is one of the important and earliest known spices produced and exported from India. It is grown in about 2,01,381 ha land with an annual production of 48,000 tonnes, largely distributed in Kerala, Karnataka accounting 92% of production, the rest being Tamil Nadu, Andhra Pradesh and North eastern states especially Assam.

India accounts for 54 per cent of the total area under pepper in the world but its share of production is only 17 per cent whereas the other countries like Brazil, Indonesia, and Malaysia account for lesser percentage of area but with more shares in the total production due to their higher productivity. Annually, India is exporting about 33 % of our production (16,000 tonnes) of black pepper earning a foreign exchange of Rs. 67,256 lakhs.

Botany

It is a climbing evergreen plant and grows to a height of 10 m or more. The vines branch horizontally from the nodes and do not attain length, but the fully grown vines completely cover the standard presenting the appearance of bush. Based on growth habits, morphological characters and biological functions, five distinct types of stem portions can be identified in the shoot system of a pepper vine (Fig. 2).

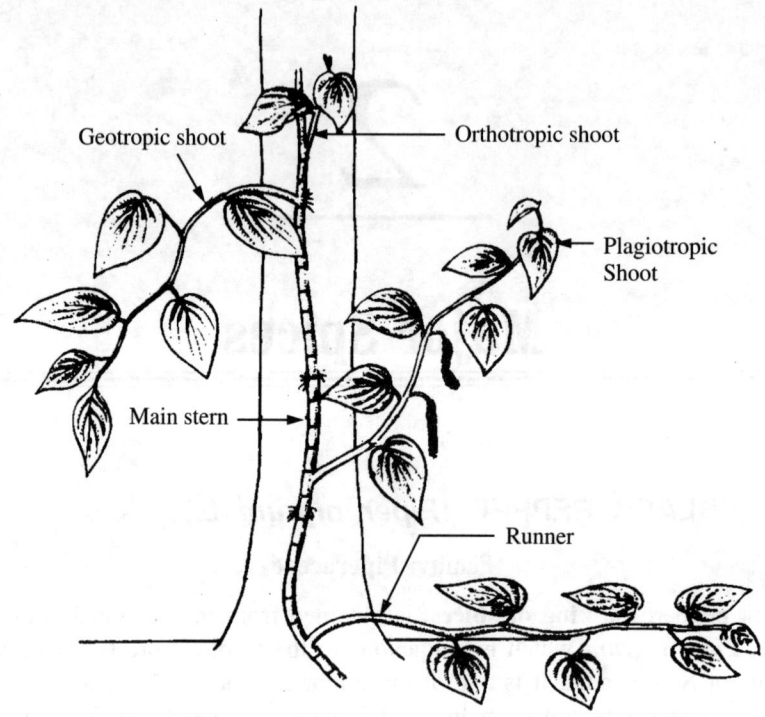

Figure 2   Types of shoots in black pepper

1. **Main stems** which originates from a seed or from a stem cutting. It climbs on a support with the aid of aerial or adventitious roots.

2. **Runner shoots** are produced from the basal portion of the main stem, growing at right angle to the main stem, usually restricted up to 50 cm from the ground.

3. **Fruiting branches (Plagiotropes)** are produced from the nodes of the main stem and they grow laterally more or less at right angles to the main stem, bearing the spikes.

4. **Topshoots (orthotropes)** - After a period of vertical growth, the top portion of the main shoots attain a bushy appearance with shorter, thicker internodes and profuse branching with large number of

adventitious roots at the nodes. This portion of the main shoot is called top shoots or orthotropes.

5. **Hanging shoots (geotropes)** - In a fully grown vine, some of the plagiotropes at the top portion are seen to give rise to a special type of shoots which hang down and grow geotropically.

The leaves are broadly lanceolate, but wide variations occur in leaf shape and are arranged alternately. The inflorescence is a catkin produced at the nodes opposite to the upper leaves. Flowers are very minute. Monoecious or dioecious or hermophrodite forms occur in different varieties. High yielding forms should have more percentage of bisexual flowers and in cultivated varieties these flowers will be more than 80 per cent. In case, if it is less, it is compensated by the higher per cent of female flowers. The male flowers are very few, 1 to 19 percent in different varieties. The fruit is a single seeded berry, which has a thin, soft pericarp surrounding the seed. It takes approximately six months to mature after flowering. Fruit setting depends upon the sex of the vine; season etc. and it will be normally about 50 percent in cultivated varieties. Sometimes, spike shedding occurs to the extent of 14 to 65 per cent causing considerable loss. Spraying of IAA 50 ppm or planofix 50 ppm and or zinc 0.5 per cent at the time of berry setting stage reduces spike shedding.

Climate and Soil

Pepper is a plant of humid tropics requiring adequate rainfall and humidity. The hot and humid climate of submountaneous tracts of Western Ghats and Eastern Ghats is ideal for its growth. It grows successfully between 20° north and south latitude and from sea level up to 1500 metres above MSL. The crop tolerates temperatures between 10° and 40°C. A well distributed annual rainfall of 125-200 cm is considered ideal for pepper.

Pepper can be grown in a wide range of soils such as clay loam, red loam, sandy loam and lateritic soils with a P^H of 4.5 to 6.0, though in its natural habitat, it thrives best on virgin soil rich in organic matter.

Varieties

Majority of the cultivated types of pepper are monoecious. Over 75 cultivars of pepper are being cultivated in India. Karimunda is the most popular of all the established cultivars of pepper among the growers of Kerala. The other important cultivars are Kottanadan, Narayakkodi, Aim-piriyan, Neelamundi, Kuthiravally, Balancotta, Kalluvally, Malligesara and Uddagare. The important characters of these popular cultivars are:

Cultivar	Fresh mean yield (Kg/vine)	Oleoresin (%)	Quality attributes		Dry recovery (%)	Features
			Piperine (%)	Essential oil (%)		
Aimpirian	4-5	15.0	4.7	2.6	34	Good for higher elevations, good in quality, late maturing.
Arakulamunda	2	9.8	4.4	4.7	33	Moderate and regular bearer.
Balankotta	1-2	9.3	4.2	5.1	35	/moderate and irregular bearer
Karimunda	2-3	11.0	4.4	4.0	35	Suitable for all pepper growing areas, high yielder, shade tolerant.
Kalluvally	1-2	8.4-11.8	2.5-5.4	3.0	35-38	Good yielder with higher dry recovery, drought tolerant.
Kottanadan	5	17.8	6.6	2.5	34-35	High yielding, drought tolerant.
Kuthiravally	3	15.0	6.0	4.5	35	High yield, good quality.
Narayakodi	1-2	11.0	5.4	4.0	36	Moderate yielder with medium quality.
Neelamundi	2	13.9	4.6	3.3	33-34	Good yielder, tolerant to *Phytophthora* infection
Vadakkan	3	10.8	4.2	3.2	-	Medium quality and yield.

Recently a number of improved cultivars have been evolved and released for cultivation

Improved varieties of Black pepper

S. No.	Name	Parentage	Yield Per vine (green berries in Kg.)	Other attributes
1.	Panniyur-1	Fl hybrid between Uthirankottah X Cheriyakania Kadan	2.5	1200 kg/ha, more adaptable to open conditions, sensitive to excess shade, dry recovery: 35.3%, oleoresins:11.8%, essential oil:3.5%, piperine: 5.3%
2.	Panniyur-2 (Krishna)	Open pollinated seedlings of Balankottah.	4.5	2828 kg/ha, shade tolerant, medium sized berries, suitable for intercropping, dry recovery: 35.7%, oleoresin: 10.9%, piperine: 6.6%. essential oil 3.4%
3.	Panniyur-3 (Shima)	Fl hybrid between Uthirankottah X Cheriyakaniya kadan.	4.4	1953 kg/ha, excessive vegetative vigour, long spike, bold berries, prefers open condition,late maturing group dry recovery: 27.8%, Oleoresin: 12.6%, Piperine: 5.2% ,essential oil 3.1%
4.	Panniyur-4	Selection from Kuthiravally type II	2.3	1419 kg/ha, stable in yield, performs well even under adverse conditions, tolerant to shade, late maturity, dry recovery: 34.7%, oleoresins: 9.2%,piperine 4.4%, essential oil 2.1%
5.	Panniyur-5	O.P.progeny of Perumkodi	2.75	1110 kg/ha, suitable for intercropping in Coconut/ Arecanut gardens, long spike, shade tolerant, medium maturity, tolerant to nursery diseases, dry recovery: 35.7%, Piperine: 5.3%, oleoresins: 12.33%, essential oil: 3.80%.

6.	Panniyur 6	Clonal selection from Karimunda type III	4.0(Potential Yield 3359)	A vigorous vine. 2127 Kg/ha Suitable for open condition as well as partial shade, spike 6 – 8 cm, more number of spikes/unit area, close setting and attractive bold berries, piperine 4.9 %, oleoresin 8.27%, essential oil 1.33% and 33.0% dry recovery.
7.	Panniyur	O.P.progeny selection from Kuthiravally	1410 (Potential yield 2770)	Vigorous vine 1410 kg/ha, suitable for open and shaded conditions, very long spike (16-24 cm), high piperine content (5.6%), oleoresin 10.6%, essential oil 1.5% and 34.0% dry recovery.
8.	Sreekara	Selection from Karimundu (K.S. 14)	4.8	2352 kg/ha, tolerant to drought, suitable for higher elevation, intercropping, medium maturity, dry recovery 35%, piperine: 5.0%, Oleoresins: 13%, essential oil: 7%.
9.	Subhakara	Selection from Karimunda (K.S.27)	4.2	2677 kg/ha, wide adaptability, suitable for intercropping as well as higher elevation, high quality, medium maturity, dry recovery: 35%, piperine; 3.4%, oleoresins: 12%, essential oil: 6%.
10.	Panchami	Selection from Aimpiriyan coll. 856	5.2	8320 kg green pepper/ha, High yielding, spikes twisted, late maturing, suitable for higher elevation, excellent fruit set. Piperine 4.7%, oleoresins: 12.5%, essential oil 3.4%, dry recovery 34.0%.
11.	Pournami	Selection from Ottaplackal type coll. No. 812.	4.7	7526 kg of green pepper/ ha, Piperine 4.1%, oleoresins 13.87, essential oil 3.4%, dry recovery 31%, tolerant to root knot nematode (*Meloidogyne incognita*) and shade, suitable

				for intercropping with arecanut and banana
12.	IISR Thevam	Clonal selection from Thevanmundi	2437 kg/ha dry berries	Stable yielding, grow vigorously, field tolerant to phytophthora, suitable for high altitude areas of South India up to 3000 ft MSL in coffee and tea estates under rainfed conditions, medium maturity Piperine 1.6%, oleoresin 8.15%, essential oil 3.1%, dry recovery 32.5%
13.	IISR Malabar Excel	F₁ hybrid between Cholamundi x Panniyur 1	1453 kg/ha dry berries	Recommended for rainfed condition, including Coffee and Tea plantation, suitable for higher elevation and plains. Piperine 2.4%, oleoresin 11.7%, essential oil 2.8%, dry recovery 32.3%.
14.	IISR Girimunda	F₁ hybrid between Narayakodi x Neelamundi	2880 kg/ha dry berries	A medium maturity group, Piperine 2.2%, oleoresin 19.65%, essential oil 3.4%, dry recovery 32.0 %.
15.	IISR Sakthi	OP Progeny of cv. Peramb-ramundi	5755 kg/ha dry berries	Tolerant to *Phytophthora* foot rot, 3.3% piperine, oleoresin 10.2%, essential oil 3.7%, dry recovery 43.0%.
16.	PLD - 2 1996	Clonal selection from Kottanandan	2475 kg/ha dry berries	Late maturity high quality cultivar, suitable for higher elevation and plains. piperine 3.0%, oleoresin 15.45%, essential oil 4.8%.

N.B. 1 to 7 released from K.A.U., Pepper Research Station, Panniyur and 8 to 15 released from IISR, Calicut, 16 released from NRC Oil Palm, Regional Station, Palode.

Propagation

Pepper is propagated by cuttings raised mainly from the runner shoots. Cuttings from the lateral branches are seldom used, since in addition to reduction in the number of fruiting shoots, the vines raised from them are generally short lived and bushy in habit. However, rooted lateral branches are useful in raising pepper in pots.

Runner shoots from high yielding and healthy vines are kept coiled on wooden pegs fixed at the base of the vine to prevent the shoots from coming in contact with soil and striking roots. The runner shoots are separated from the vine in February-March and after trimming the leaves, cuttings of 2 to 3 nodes each are planted either in nursery beds or polythene bags filled with fertile soil. Cuttings from middle 1/3 of the shoots are desirable as they are high yielding. Adequate shade is to be provided and irrigated frequently. The cuttings will strike roots and become ready for planting in May-June.

A rapid multiplication technique has been developed by the Indian Institute of Spices Research, Calicut. In this method, a trench of 0.75 m deep and 0.3m wide having convenient length is made. The trench is filled with rooting medium (preferably forest soil, sand and cowdung mixture at 1:1:1). Split halves of bamboos with septa of 8 to 10 cm diameter and 1.25 to 1.50 m length are fixed at 45° angle on a strong support (fig. 3). The bamboos can be arranged touching one another. Rooted cuttings are planted in the trench at the rate of one cutting each for one bamboo. The lower portions of the bamboo are filled with a rooting medium (coir dust and cattle manure mixture 1:1) and the growing vine is tied to the bamboo in such a way as to keep the nodes pressed into the rooting medium. The tying could be done with dried banana sheath fibre. The vines are irrigated regularly. As the vines grow up, filling up the bamboo with rooting medium and tying each node, pressing it down to the rooting medium are to be continued regularly. For rapid growth, each vine is fed at 15 days interval with 0.125 litre of nutrient solution prepared by dissolving urea (1kg), super phosphate (0.75 kg), Muriate of potash (0.5 kg) and Magnesium sulphate (0.25 kg) in 250 litres of water.

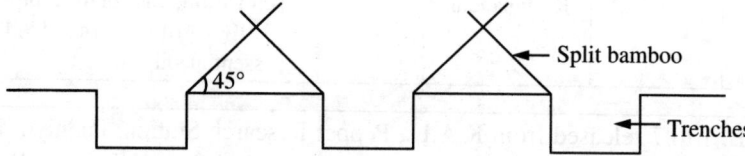

Figure 3 Arrangement of bamboos in rapid multiplication technique.

When the vine reahces the top in about 3 to 4 months, the terminal bud is nipped off and the vine is crushed at about three nodes above the base, in order to activate the axillary buds. After about 10 days, each vine is cut at the crushed point and removed from the rooting medium and each node is separated. Such cuttings with the bunch of roots intact is planted in polybags filled with pot mixture (Fig.4) and kept in a cool humid place. *Trichoderma* @ 1 kg and VAM @ 100cc / kg of soil can be added to the potting mixture.Care should be taken to keep the axil above the soil. The buds start developing in about 3 weeks when the polybags can be moved and kept in semi shade. Subsequent harvesting can be had at every 2-2½ months time. The advantages of this method are : 1) multiplication is rapid (1:40), (2) the root system is well developed, and (3) a better field establishment and more vigorous growth as a result of better root system.

IISR, Calicut has developed a trench method in which single node cuttings are made to root in polybags kept in pits of 2.0 × 1.0 × 0.5 m size and covered with polythene sheets. The cuttings are to be watered atleast five times a day with rose can and then covered. After 2-3 weeks of planting, the cuttings start producing roots and the watering frequency can be gradually reduced and hardened. The cuttings can be taken out of the pit within two months.

In Vietnam, orthotropic shoots are used for propagation. These shoots are cut or extracted from top up to 6-7 nodes down on the support. Too tender and two woody shoots need to be avoided. The medium mature green shoots are selected and cut into bits of single node and planted in the poly bags containing nursery mixture. They are housed in humid poly houses to induce rooting. Rooting success varies from 70-80 percent. This has many advantages such as (1) totally free from soil contamination, (2) Flowering is early and strats fruiting by 2nd year and (3) Fruiting laterals start right from the base which is cylindrical in shape, giving higher yield. The main drawback is the limited availability of the orthotropic shoots.

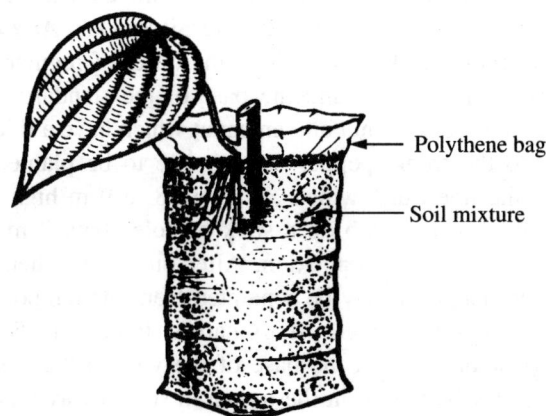

Polythene bag

Soil mixture

Figure 4 Single node cutting separated from rapid multiplication technique.

Selection of site

Well drained level land and hill slopes are suitable for growing pepper. When grown on a slopy land, the slopes facing south should be avoided and the lower half of north and north eastern slopes are preferred for planting; so that the vines are not subjected to the scorching effect of the sun during summer.

System of cultivation

Pepper is grown as monocrop as well as a mixed crop. Large scale cultivation of pepper as monocrop is done on hill slopes by clearing jungle lands and planting standards for the vines to climb on. As a mixed crop, it is grown with arecanut, coconut, mango, jack etc. where these trees serve as standards for the pepper vines. Pepper is also a suitable intercrop in coffee estates where the shade trees serve as good standards for them.

Planting

Pepper cuttings are generally planted with the onset of the south west monsoon. Planting can also be done during the north east monsoon wherever it is regular and well distributed. When pepper is grown as pure crop, pits of 0.5 m cube are dug at a spacing of 2.5 × 2.5 m and *Erythrina* stem cuttings of 2 m length or its two year old seedlings are planted on receipt of early monsoon showers. The pits are filled with a mixture of top soil, farmyard manure @ 5 kg / pit and 150 g rock phosphate. Neem cake @ 1 kg and Trichoderma harzianum @ 50 g also may be mixed with the mixture at the time of planting. Certain other trees like Silver Oak, *Ailanthus excelsa* and *Garuga pinnata* are also used. With the onset of regular rains, 2 or 3 rooted cuttings are planted around the base of the standard nearly 30 cm away. But in the case of coconut and arecanut which have a thick intercoiled root net close to the trunk, pepper cuttings are to be planted 100 to 120 cm away from the tree trunk which are about 8 to 9 m high. Initially, the vines may be allowed to climb on a stick or pole about 2 m tall which is tied to the trunk in a slanting position. After one year, when the vine has attained sufficient length it may be separated from the temporary stake and the lower leaves may be nipped off. A narrow trench of 15 cm deep and wide should be prepared from the base of the vine to the base of the tree trunk. The vine may be placed in the trench in such a way that the growing tips are tied to the trunk while the other part of vine is covered with the soil. A small ridge is formed over the trench which should not be disturbed while doing intercultural operations to the palm.

Cultural Practices

As the cuttings grow, the shoots are tied to the standards as often as required. The young vines should be protected from hot sun during summer months by providing them with artificial shade. Regulation of shade by lopping the branches of standards is necessary not only for providing optimum light to the vines but also for enabling the standards to grow straight. Adequate mulch with green leaf saw dust or coir dust or organic matter should be given towards the end of north-east monsoon. The base of the vines should not be disturbed to avoid root damage. During the second year, practically the same cultural practices are repeated. However, lopping of the standards should be done carefully from the fourth year onwards, not only to regulate the height of the standards, but also to shade the pepper vines optimally. Excessive shading during flowering and fruiting encourages pest infestation. Pruning the top of the vine after it has reached the required height i.e., 6 m is normally practiced when it is trained on standards like silver oak, coconut, arecanut for convenience of picking.

From the fourth year, usually two diggings are given, one during May-June and the other towards the end of south-west monsoon in October-November. Growing cover crops like *Calapogonium mucanoides, Mimosa invisa* are also recommended under west coast conditions. Young pepper vines need to irrigated regularly in summer months

Crop failure due to 'spike shedding' is becoming production constraint in many areas. It is attributed to various factors *viz.,* fungal (*Colletotrichum sp.*), insects, drought, absence of pollination and inadequate pollination. Predominance of female flowers in the spikes is reported as the major cause for pollination failure and resulted in spike shedding. Light availability also plays an important role in pollination which has a direct bearing on productivity of black pepper. A minimum of 250-400 mm of rainfall during March to May is required in shade regulated pepper gardens for early spiking and setting followed by sufficient rainfall during June for elongation of spikes. Basal irrigation of pepper vine during March – May and shade regulation in April helps in early initiation of spikes and good setting. Black pepper may be irrigated during summer months in year of pre monsoon failure to harvest good crop in high altitude in coffee based cropping system where black pepper is grown as mixed crop. Yielding pepper vines need to be irrigated from April first week to May last week by sprinkler or hose or pot to the plant base. Each vine needs 50-60 liter of water in an interval of 10-12 days, providing totally around 250-300 liter per vines during the dry months. This leads to early initiation of spike, good setting, limited anthracnose incidence and more number of bisexual flower and good yield. Besides annually basin management by applying tank silt to height of 2-3 cm thickness is desirable.

Manuring

Judicious and regular manuring is necessary to get good yield. About 10 kg of well rotten cattle manure or compost is given in April-May. Fertilizers to supply 100 gN, 40 g P_2O_5 and 140 g F_2O per standard for vines of three years and above may be applied annually in two split doses in April-May and August-September. During the first year of planting, 1/3 of the above dose and in the second year 2/3 of the above dose may be given. Manures are applied around the vines at a distance of 30 cm and forked into the soil. Lime may be applied at the rate of 500 g per standard during April in alternate years. When biofertilizer like *Azospirillum* is applied @ 100 g / vine, the recommended nitrogen dose may be reduced by half to 70 g/ vine. In soils that are deficient in zinc or magnesium, foliar application of 0.25% zinc sulphate twice a year (May-June and September- October) and soil application of 150 g/ vine magnesium sulphate, respectively is recommended.

Harvesting and curing

Pepper vines start yielding usually from the 3rd or 4th year. The vines flower in May-June. It takes 6 to 8 months from flowering to ripening stage. Harvesting is done from the November to February in the plains and January to March in the hills. When one or two berries on the spike turn bright or red, the whole spike is plucked.

Berries are separated from the spikes by rubbing them between the hands or trampling them under the feet. After the separation, the berries are dried in the sun for 7 to 10 days until the outer skin becomes black and shrunken and assumes the characteristic wrinkled appearance of commercial black pepper.

For making good quality black pepper of uniform colour, the separated berries are collected in a perforated bamboo basket or vessel and the basket with the berries is dipped in boiling water for one minute. The basket is then taken out and drained. The treated berries are sun dried on a clean bamboo mat or cement floor. The recommended drying surfaces are bamboo mat coated with fenugreek paste, cement floor and high density black polythene which gives better appearance and cleanliness to the dried proc\ duct. Mechanical driers such as copra drier, convection drier and cascade type driers can also be employed for drying. The optimum temperature to be maintained in mechanical driers should be around 60° C.

White pepper of commerce is prepared by removing the outer skin and the pulp below it before drying the berries. Spikes with fully ripe berries are filled in gunny bags and steeped in flowing water for about 7 days. Outer rind of the berries is then removed by rubbing them with hands in a bucket of water and further cleaning the seeds with fresh water. The cleaned seeds are dried for 3 to 4 days. The seeds which are now dull white in colour are further cleaned by winnowing and polishing them by rubbing with a cloth. The recovery of white pepper is about 25 percent of ripe berries while that of black pepper is about 33 per cent. White pepper contains lesser volatile oil than black pepper.

The dried pepper is cleaned to get rid of the extraneous matters such as dirt, stalks, leaves etc. Magnetic separator is used to remove metallic contamination such as iron fillings and stray nails. Vibratory conveyors with inclined decks in combination of air classification are used for efficient de-stoning of spices. Broken pepper and light pepper grades are separated pneumatically; pin heads which come along with garbled pepper are separated by sieving. TNAU has developed a hand operated cleaner cum grader suitable for cleaning and grading operations. The unit consists of a rotor made of sieves, shaft. Screw auger, handle, hopper, frame, outlets and handle. Along the length of the rotor is divided into three segments of each 450 mm to mount sieves of various opening. A screw is provided inside the rotor for easy converting of the feed materials to the sieves of various opening. A screw is provided inside the rotor for easy converting of the feed materials to the sieve perforations. A feed hopper to hold about 15 kg of pepper has been provided at the feed inlet end with appropriate side slopes for easy feeding of the feed into the sieves. Four inclined outlets are provided for collection of impurities, cleaned and graded products. The unit is provided with three sieves with round holes of size, 3.5mm, and 3.8 mm and 4.8 mm diameters. These sieves are as per the Agmark specifications.

Different grades based on size are Pin heads, Light pepper, Malabar Garbled, Tellicherry Garbled EB, Tellicherry Garbled Spl.EB, Malabar ungarbled ,Tellicherry ungarbled, High range ungarbled and Half pepper.

Considerable advances have been made in recent years in the diversification of value added processed products from pepper which have great demand. They include 3 major groups viz., (a) green pepper based products - canned or bottled green pepper in brines, cured green pepper, frozen green pepper, freeze dried green pepper, dehydrated green pepper, green pepper pickles, green pepper flavoured products white pepper (whole) or powder etc., b) black pepper based products - Black pepper powder, pepper oleoresins, pepper oils etc and (c) pepper by-products which have medicinal, culinary and industrial uses. These processed products earn more foreign exchange per unit weight/volume.

Yield

Pepper vines attain full bearing stage in the 7th or 8th year after planting and it starts declining after 20 to 25 years and replanting has to be done thereafter. One hectare plantation of 7 or 8 years old gives about 800 to 1000 kg of black pepper. Well maintained plantations in coffee based system under irrigation management system can yield up to 10 kg of dry pepper per vine.

Bush pepper

It is a method of cultivating the vine in the form of a bush. One year old healthy fruiting branches are selected with 3 to 5 nodes and all the leaves except the flag leaf are removed and planted in a shaded area in the nursery, either in trenches or in polybags (45 × 30 cm) containing moist coir dust. Before planting, the cuttings are dipped in 1000 ppm of IBA for 45 seconds. After planting, the trenches are covered with polythene sheets and in the case of polybags; the mouth is tightly tied with coir thread to avoid moisture loss. They normally root in 30 to 50 days. Such rooted cuttings are planted in pots or fields after sufficient hardening treatment. Cuttings grow like a bush and flower in the same year itself. These bushes produce more and more of fruiting branches only. Adequate manuring i.e., 2 to 5 kg of FYM along with 10 g of NPK 1:1:2 mixtures may be given per bush at 3 months interval. Watering and plant protection may be adopted according to necessity. Under average management a good bush pepper plant may yield 1.5 kg green pepper in a span of 2 to 3 years.

Plant protection

The important pests and diseases in pepper, their symptoms/damage and control measures are furnished below:

Pest/Disease	Symptoms	Control measures
Pests Pollu beetle (***Longitarsus,*** ***nigripennis***)	Adults feed on growing points, tender shoots, leaves, tender spikes and berries. Grubs bore into the berries and cause	Regulation of shade in the plantation reduces the population of the pest in the field. Spraying quinalphos (0.05%) during June – July and September – October and Neemgold (0.6%) (Neem –

	black colour and they crumble when pressed.	based insecticide) durin_ August, September and October is effective for the management of the pest. The underside of leaves (where adults are generally seen) and spikes are to be sprayed thoroughly.
Top shoot borer *(Cydia hemidoxa)*	Caterpillars bore into the tender. Shoots turning them to black and drying up.	Spray quinalphos 0.05% twice when new shoots emerge.
Leaf gall thrips *(Liothrips karnyi)*	Thrips live in colonies within the tubular marginal galls induced by them. Such leaves become thick malformed, crinkled.	Spray dimethoate (0.05%) during new flush emergence.
Scale insects *(Lepidosaphes piperis)* *(Aspidiotus destructor)*	Drying of the infested portion of the vine.	Spray dimethoate 0.05% at 15 days intervals.
Diseases Quick wilt or foot rot *(Phytophthora capsici)*	Pathogen infects leaves, spikes, collar region and roots. Collar infection causes sudden collapse of the vine, other symptoms is foliar yellowing and defoliation. High soil moisture and relative humidity during monsoon favour infection and spread.	Removal of the infected dead vines and burn them, provide proper drainage; avoid damage to root system and stem of vines. After the receipt of a few monsoon showers (May-June), all the vines are to be drenched at a radius of 45-50 cm with copper oxychloride 0.2%@ 5-10 litres/vine. A foliar spray with Bordeaux mixture 1% is also to be given. Drenching and spraying are to be repeated once again during August-September. A third round of drenching may be given during October if the monsoon is prolonged. After the receipt of a few monsoon showers, all the vines are to be drenched with potassium phosphonate 0.3% is also to be given. A second drenching may be given during October.

		After the receipt of a few monsoon showers, all the vines are to be drenched with 0.125% metalaxyl mancozeb @ 5-10 litres / vine. A foliar spray with metalyaxl mancozeb 0.125% may also be given. At the onset of monsoon (May-June), apply *Trichoderma* around the base of the vine @ 50g/vine. A foliar spray with potassium phosphonate 0.3% or Bordeaux mixture 1% is also to be given. A second application of same composition is to be given during Aug – Sep.
Pollu disease *(Colletotrichum gloeosporioides)*	Brown sunken patches seen in the young berries, develops characteristic cross splitting and finally turn to black in colour and dry.	Spray 1% Bordeaux mixture.
Slow decline or Slow wilt (Soil borne fungi *Fusarium sp., Rhizoctonia sp., Pythium sp. Diplodea sp. Nematodes-Radopholus similis Meloidogyne incognita*)	During dry months, foliar yellowing defoliation and dieback symptoms appear in vines and during monsoon some of the affected vines recover and put forth fresh foliage and again they show decline symptoms during dry months, thus gradually loosing vigour and productivity.	Remove the severely affected vines, treat the planting pit with phorate @15g or carbofuran @50g at the time of planting, apply phorate @30g or carbofuran @100g/vine twice in a year. Biocontrol agents like *Pochonia chlamydosporia* or Trichoderma harzianum can be applied @ 50 g / vine twice a year (during April – May and September – October).
Stunt Disease (Viruses)	The vine exhibit shortening of internodes to varying degrees. The leaves become small and narrow with varying degrees of deformation and appear leathery,	Use virus free healthy planting material, regular inspection and removal of infected plants; the removed plants may be burnt or buried deep in soil. Insects such as aphids and mealy bugs on the plant or standards

puckered and crinkled. Chlorotic spots and streaks also appear on the leaves occasionally. The yield of the affected vines decreases gradually. Two viruses namely Cucumber mosaic virus and a Badnavirus are associated with the disease. The major means of spread of the virus is through the use of infected stem cuttings. The disease can also be transmitted through insects like aphids and mealy bugs.	should be controlled with insecticide spray such as dimethoate or monochrotophos @ 0.05%.

2.2. CARDAMOM (*Elettaria cardamomum* (L.) Maton)

Family: **Zingiberaceae**

Cardamom, popularly, known as **Queen of Spices** is native to the evergreen rainy forests of Western Ghats in South India. It is cultivated in about 71,170 ha with an annual production of 19,380 tonnes, mainly confined to the Southern States viz., Kerala, Sikkim, Karnataka and Tamil Nadu accounting for about 28% world production. Nearly 40 per cent of our production is exported to more than 60 countries earning a foreign exchange of nearly 13216 lakhs rupees. Cardamom is used for flavouring various preparations of food, confectionary, beverages and liquors.

Botany

Cardamom is an herbaceous perennial having underground rhizomes. The aerial pseudostem is made of leaf sheaths. Inflorescence is a long panicle with racemose clusters arising from the underground stem, but comes up above the soil. Flowers are bisexual, fragrant, fruit is a trilocular capsule. Flower initiation takes place in March - April and from initiation to full bloom, it takes nearly 30 days and from bloom to maturity, it takes about 5 to 6 months.

Honey bee is the principal pollinating agent and it increases the fruitset considerably when compared to flowers prevented from bee visits. Cardamom flowers remain open for 15 to 18 hours and stigma receptivity and pollen viability are maximum during morning hours. Four to five bee hives per ha should be maintained and pesticide spraying should be carefully monitored to avoid any damage to the bees. Large cardamom or Nepal cardamom or Greater Indian cardamom is the dried fruits of *Amomum subulatum.* It is native of Eastern Himalayan region and is now cultivated in Sikkim, Darjeeling and Assam hills in about 29000 ha in India with an annual production of 5640 tonnes. About 775 tonnes are annually exported earning around Rs.4462 lakhs.

It is a perennial crop, propagated from the seeds or cut bits of the rhizome. It starts bearing in 3 to 5 years after planting and the economic age of the plantation is 12 to 15 years. The fruits are about 2.5 cm long, ovoid and triangular in shape brown or pink in colour when ripe. They contain 40 to 50 seeds. Average yield is 300 to 1000 kg per ha from 4th or 5th year.

Climate and Soil

The natural habitat of cardamom is the evergreen forests of Western Ghats. It is grown in the areas where the annual rainfall ranges from 1500 to 4000 mm, with a temperature range of 10 to 35°C and an altitude of 600 to 1200 m above MSL. Rainfall distribution should be good and summer showers during February-April are essential for panicle initiation, otherwise it will affect the yield. With the denudation of forests in the Western and Eastern Ghats, the favourable ecosystem has been affected destabilising the microclimate and rainfall in the cardamom growing tracts, resulting in poor growth and yield. Cardamom grows luxuriantly in forest loamy soils, which are generally acidic in nature with a pH range of 5.5 – 6.5. Growth of cardamom is enhanced, when planted in humus rich soils with low to medium available phosphorous and medium to high available potassium.

Varieties

Based on the size of the fruit, two varieties are broadly recognised viz., *Elettaria cardamomum* var. *major* consisting of wild indigenous types and var. *minor* comprising the cultivated types viz., Mysore, Malabar and Vazhukka (natural hybrid between Mysore and Malabar). These types are identified mainly based on the nature of panicle and shape and size of fruits as follows:

Particulars	Mysore type	Malabar type	Vazhukka
Plant stature Panicle Capsule Adaptability	Robust Erect Bold, elongated Higher altitudes (900-1200m MSL) in Kerala	Medium sized Prostrate Round or oblong Lower altitudes (600- 900m MSL)	Robust Semi-erect Round to oblong Wide range

Recently a number of improved cultivars have been released for cultivation.

S. No.	Varietal Name	Type	Yield Potential (dry capsules/ ha in kg.)	Other attributes
1.	Mudigere- 1	Clonal selection from Malabar type	275(Potential yield 1000)	Erect and compact plant, suitable for high density planting, moderately tolerant to thrips hairy caterpillar and white grubs, oval bold capsules, contained 8.0 % oil, 36.0%, 1, 8 cineol 42.0%, α-terpenyl acetate, dry recovery 20.0%.
2.	Muidgere- 2	Clonal selection from open pollination of Malabar type	475 (Potential yield 1000)	Early maturing, suitable for high density planting, round/oval and bold capsules, oil 8.0%, 1, 8 cineol 45.0%, α-terpenyl acetate 38.0%.
3.	PV 1	A selection from Walayar collection, a Malabar types	260 (Potential Yield 500)	An early maturing types, short panicle, elongated slightly ribbed light green capsules, essential oil 6.8%, 1, 8 cineol 33.0%, α-terpenyl acetate 46.0%, dry recovery 19.9%.
4.	PV 2	A selection from O.P. seedling of PV -1, a Malabar type	982 (Pot. Yield 1250)	Early maturing, unbranched lengthy panicle, long bold capsules, high dry recovery percentage

				(23.8%), essential oil 10.45%, field tolerant to stem borer and thrips. Suitable for elevation range of 1000 to 1200 meters above MSL.
5.	ICRI 1	Selection from Chakkupallam collection a Malabar type	325 (656 kg under irrigation condition)	An early maturing variety, medium sized panicle with globose, round and extra bold dark green capsules; contains oil 8.7%, 1, 8 cineol 29.0%, α-terpenyl acetate 38.0%, dry recovery 22.9%.
6.	ICRI 2	Clonal selection from germplasm collection, Mysore type	375 (766 kg under irrigated condition)	Performs well under high altitude and irrigated condition, medium long panicles, oblong, bold and parrot green capsules, tolerant to azukkal diseases. Dry recovery 22.5%
7.	ICRI 3	Selection from Malabar	440 (790 kg under irrigated condition	Early maturing, non-pubescent leaves, tolerant to rhizome rot disease, oblong, bold parrot green capsules, oil 6.6%, 1 , 8% cineol 54.0%, α-terpenyl acetate 24.0%, dry recovery 22.0% suitable for hill zone of Karnataka.
8.	TDK 4	Clonal selection from Vadagaraparai area of lower Pulenys, a Malabar type	455(960 kg under irrigated condition)	Early maturity, medium sized panicle, globose bold capsules, oil 6.4%. Suitable for low rainfall areas, relatively tolerant to rhizome rot and capsule borer
9.	IISR Survasini	Selection from OP progeny of CL 37 from RRS Mudigere,	745(1322 kg under irrigated condition)	Early maturity, suitable for high density planting, long panicle, high % of bold parrot green

				capsule (89), essential oil 8.7%, 1, 8 cineol 42.0%, α-terpenyl acetate 37.0%,dry recovery 22.0% ; tolerant to rhizome rot, thrips, shoot/ panicle/capsule borer.
10.	IISR Vijetha	Clonal selection from field resistant plant for katte, a Malabar type	643 (Potential yield 979)	Virus resistant selection with high percentage of bold capsules. (77.0%) oil 7.9%, 1, 8 cineol 45.0%, α-terpenyl acetate 23.4%, dry recovery 22.0%. Recommended to moderate rainfall areas with moderate high shaded and mosaic infected areas, field tolerant to thrips and borer as well as mosaic.
11.	IISR Avinash	A selection from OP seedlings of CCS-1, a Malabar type	847 (Potential Yield 1483)	Has extended flowering period, dark green capsules and retains its colour even after processing oil 6.7 %, 1, 8 cineol 30.4%, α-terpenyl acetate 35.5%, dry recovery 22.8%. Tolerant to rhizome rot prone areas and intensive cultivation.
12.	Njallani Green Gold	A Clonal selection from Vazhukka type	1600 (Potential Yield > 3000)	A high yielding clonal selection by a cardamom grower in Idukki district. Capsules bold, over 70% of the cured cardamom above 7mm. Reported to be high quality. Currently the highest yielding cultivar. Yield up to 5000 kg/ha has been recorded.

N.B. 1 and 2 released from RRS, UAS, Mudigere, 3and 4 released from Cardamom Research Station, Pampadumpara, 5 to 7 released from ICRI, Mylamdumpara, 8 released from ICRI, Thadiyankudisai, 9 to 11 released from IISR, Cardamom Research Center, Appangala, 12 released from farmers selection from Idukki.

Nursery

Cardamom is propagated mainly through seeds and also through suckers each consisting of at least one old and a young aerial shoot. The suckers are commonly used for gap filling but suckers may not be available in larger numbers. Therefore, a rapid clonal multiplication technique evolved by IISR, Cardamom Research Centre, Appangala, is proved to be quick, reliable and economic for production of large number of quality planting materials. The site selected for this method should have a gentle slope and must be nearer to the water source. Trenches of 45 cm width, 45 cm depth and of any convenient length may be taken across the slope or along the contour at 1.8 m apart. The top 20 cm depth soil is excavated separately and heaped on the upper side of the trench. The lower 25 cm depth soil is excavated and heaped on lower side of the trenches all along the line. The top soil is mixed with equal portions of humus rich jungle soil, sand and cattle manure and filled back by leaving a depression of 5 cm at the top to facilitate mulching for retention of soil mixture. Suckers, each consisting of one grown up tiller and a growing young shoot, (Fig. 5), are placed at a spacing of 0.6 m in the trenches during March-October. Regular cultural operations are to be followed including a high fertilizer dose of 100:50:200 kg NPK/ha in 6 split doses at 60 days interval along with neem cake at 250 g/plant. Irrigation should be provided at least twice a week. Over head pandal at a height of 3.6m covered with coir mat or leafy twigs of any shade tree may be provided during the non-rainy season. Within a period of 12 months, a plant would produce atleast 32 to 42 suckers which may yield at least 16 to 21 planting units, ie., about 1.5 lakhs planting units per ha. of clonal nursery within 12 months of planting.

Seedlings are normally raised in primary and secondary nurseries. The nursery site should be selected on gentle slopy lands, having an easy access to a water source. Raised beds are prepared after digging the lands to a depth of 30-45 cm. The beds of 1 m width and convenient length raised to a height of about 30 cm are prepared. A fine layer of humus-rich forest soil is spread over the beds. Seeds are to be collected from well ripe capsules. Immediately after harvesting, the husk is removed and the seeds are washed repeatedly in water for removing the mucilaginous coating. After draining the water, the seeds are to be mixed with wood ash and dried in shade for a day. In order to ensure uniform and early germination, seeds should be sown immediately after extraction. If the sowing is delayed, pre-sowing treatment of seeds with 25% Nitric acid for 10 minutes is advisable to get a quick and higher germination. In addition, soaking the seeds in solutions

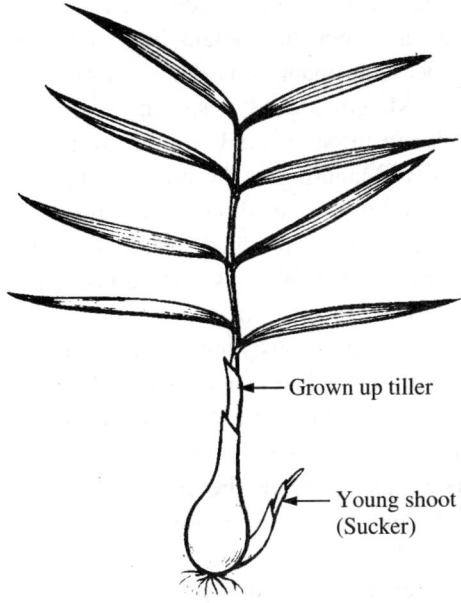

Figure 5 A planting unit in cardamom

of gibberelic acid (GA$_3$) and ethrel enhances germination. One kg of seed capsules may produce 5000 seedlings.

Sowing may be taken up during November-January and is done in rows spaced at 10 cm and 1-2 cm apart within the row and the seed rate for 6x1 m mixed bed is 30-50 g. deep sowing of seeds has to be avoided for better and quick germination. Seed beds are to be dusted with BHC10%. Beds are mulched to a thickness of 2 cm with paddy straw or any locally available material and are watered regularly. The germination commences in about 30 days and may continue for a month or two. After germination, the mulch is to be removed.

An overhead pandal with a height of 2 m is quite desirable. Materials like coir mat, plaited coconut leaves or tree twigs which do not shed their leaves easily may be used but the coir mat is preferred as it allows uniform filtered sunlight.

The excess seedlings are to be thinned out after 75-80 days of sowing. The thinned out seedlings may be used for gap filling within the nursery bed or for raising secondary nursery. When the seedlings attain 5-6 leaf stage, light earthing up is to be done. This would encourage better tillering and proper growth of seedlings.

Generally in Kerala and Tamil Nadu, the seedlings are transplanted to the secondary nursery when they attain four to six leaf stages. The beds are prepared in the same manner as that of a primary nursery. Seedlings are transplanted in March- May at a spacing of 20 x 20 cm and mulched immediately. Beds are to be covered with an over head pandal and should be watered regularly. Manuring at the rate of 90 g N, 60 g P_2O_5 and 120 g K_2O per bed of 5xlm size, in three equal split doses at an interval of 45 days is recommended to produce healthier seedlings. The first dose of fertilizer may be applied 30 days after transplanting in the secondary nursery. Polythene bags of size 20 x 20 cm and 100 gauge thickness are filled with potting mixture consisting of forest top soil, cow dung and sand(ratio 3:1:1). Sufficient holes at the base of polybags are provided to ensure good drainage. Seedlings at 3-4 leaf stages are transplanted into each bag (one seedling/bag). Seedlings raised in the polybags have a uniform growth and nursery period could be reduced by 5-6 months. Recently, instead of secondary nursery beds, the seedlings are also raised in polybags containing rich forest soil.

In Karnataka, ten months old seedlings are used for planting in the main yield, while in Kerala and Tamil Nadu, 18 months old seedings are commonly used.

Preparation of land: All undergrowth should be cleared and excess shade trees or branches should be thinned out to have an even overhead canopy. In open areas like marshy valleys and grasslands, shade trees are to be raised before planting cardamom seedlings. Fast growing shade trees like Dadap (Erythrina *lithosperma*), Albizia, Karuna (Vernonia *arbores*), Corangati (*Acrocarpus fraxinifolius*), Chandana Vembu (*Toona ciliata*), Njaval (*Syzygium cumini*), Jack tree (*Atrocarpus heterophyllus*) etc. are recommended to protect the seedlings from direct sunlight. Pits of 45x45x30 cm size are dug in April- May and filled with a mixture of top soil and compost or well decomposed farm yard manure. In slopy land, contour terraces may be made and pits may be taken along the contour and a close planting (2mxlm) is advisable along the contour. The spacing adopted in Karnataka for the Malabar type is 2x2 m between plants and rows. In Kerala region 2-3 m on either side is adopted. Staggered trenches may be taken across the slope to conserve run off rain water. The soil collected in trenches may be utilised for earthing up during the post-monsoon period.

Planting: The planting is carried out during the rainy season commencing from June. Under Eastern Ghat hills, July planting is adopted. Seedlings are to be planted upto the collar region for better growth. Cloudy days with light drizzle are ideal for planting.

10 to 18 month old cardamom seedlings are selected for planting in the pits. While planting, 15 g of carbofuran (Banned in Kerala) or 50 g neem cake and rock phosphate (40 g) are applied to the pit. Deep planting should be avoided, as it results in suppression of the growth of new shoots and might result in death of the plants. Stakes may be provided to avoid the damage caused by wind and the plant base need to be covered with suitable mulching material.

Planting diagonally to the slopes helps to prevent run off. Trench system of planting (60 × 30 cm) with a spacing of 2 × 1 m is generally preferred over pit system, as it results in better establishment of the plants, higher yield and greater moisture retention. In sloppy lands, contour terraces need to be prepared and pits are taken along the contours at 2 × 1 m spacing. Based on slope, terraces are made at 2-3 m between the contours.

For Mysore and Vazhukka cultivars, plant to plant distance can be 3 x 3 m (1111 plants / hectare) and 2.4 x 2.4 m (1736 plants / hectare) respectively. A spacing of 1.8 x 1.8 m or 2.0 x 2.0 m is ideal for Malabar types in Karnataka (2500 -3000 plants / ha)

In some parts of tea estates in South India, redgum (*Eucalyptus* sp.) is planted in flat valley bottoms at a regular spacing. Cardamom underplanted in this redgum area produces better growth as equal to natural jungle areas.

Recently, mixed cropping of cardamom in arecanut, rubber and coffee plantations is gaining an impetus especially with small growers which assure greater significance in the light of frequent dry spells and fluctuating price structure.

Mulching: It is an important cultural practice in cardamom. Fallen leaves of the shade trees are utilised for mulching. Sufficient mulch should be applied during November - December to reduce the ill effects of drought which prevails for nearly 4 to 5 months during summer. Exposing the panicle over the mulch is beneficial for pollination by the bees.

Weeding: Cardamom being a surface feeder in the first year of planting itself, weeding at frequent intervals is necessary. Depending upon the intensity of weeds, 2 to 3 rounds of weeding are necessary in a year. The first round of weeding is to be carried in May-June, the second in August - September and the third in December -January. In slopy land slashing of weeds is alone to be carried out otherwise it encourages to more soil erosion. Weedicides like paraquat @ 625 ml in 500 litres of water may be sprayed in the interspaces between rows leaving 60 cm around the plant base.

Trashing: Trashing consists of removing old and drying shoots of the plant once in a year with the onset of monsoon under rainfed conditions and 2-3 times in high density plantations provided with irrigation facilities. Trashing from November onwards may be avoided due to the ensuing dry season.

Shade regulation: Cardamom being a pseophyte is very sensitive to moisture stress. Shade helps to regulate soil moisture as well as temperature and provides congenial micro-climate for cardamom. Excess shade is also quite detrimental and shade has to be regulated so as to provide 50-60 percent filtered sunlight. Cardamom plants can tolerate less shade in areas where well distributed and adequate rainfall is received. In Guatemala, it is practically grown in open, since rainfall is received round the year. In South India, many trees are available in the natural habitat to provide shade but an ideal shade tree should have a wider canopy, minimum side branching and it should not shed the leaves during flowering phase of Cardamom so as not to affect pollination. Some of the common shade trees in Cardamom estates are Palangi *(Artocarpus fraxinifolius)*, Jack, Red cedar *(Cedrella toona)*, Karimaram *(Diospyros ebenum)* and Elangi *(Mimusops elangi)*. The temporary shade trees like *Erythrina lithosperma* and *E.indica* are most unsuitable as they compete for nutrients and soil moisture. In order to provide adequate light during monsoon, shade regulation may be taken up before the onset of monsoon. A two tier canopy with a height of not more than 3 m between the lower and higher canopy may be maintained. Areas exposed to western side should have adequate shade.

Earthing up: After the monsoon is over, a thin layer of fresh fertile soil, rich in organic matter may be earthed up at the base of the clump, covering up to the collar region by scraping between the rows or collecting soil from staggered trenches/check pits. This encourages new growth.

Irrigation: In order to overcome the dry spell during summer, it is necessary to irrigate the crop to get maximum production as it helps in initiation of panicles, flowering and fruitset. Depending on the moisture holding capacity of soil and topography of the estates, they may be irrigated at an interval of 10 to 15 days till the onset of monsoon. Sprinkler irrigation and or drip irrigation at the rate of 4 litres per clump per day during dry months increases the yield. Drip irrigation helps to ensure growth round the year and the fruits are available for picking almost 8 -10 months in a year.

Manuring: A fertilizer does of 75 kgN, 75 kg P_2O_s and 150 kg K_2O per ha is recommended under irrigated condition for high yielding plantations yielding 100 kg/ha and above and a dose of 30:60:30 kg/ha is recommended for gardens under rainfed condition. Besides, organic manures like compost or cattle manure may be given @ 5 kg per clump. Neem oil cake may also be applied @ one kg per clump. Foliar application of urea at 3%, single super phosphate at 1%, and muriate of potash at 2% is also beneficial to imporve the yield. In places where zinc deficiency is noticed, spraying of zinc 500 ppm twice in a year, April-May and September-October is recommended. Soil application of boron @ 7.5 kg/ha in two splits along with NPK is also recommended.

Fertiliser is applied in two split doses. The first application during May will help in the production of suckers and development of capsules and the second application during late September will help the initiation of panicles and sucker. Only half the dose of fertilizer is to be applied during the first year and full dose is given from second year onwards.

Being a surface feeder, deep placement of fertiliser is not advocated. Appiication of fertiliser is done at a radius of 30 cm and covered with a thin layer of soil.

The principal pollinating agent in cardamom is honey bee (*Apis ceranaindica*). Maintaining four bee colonies per hectare during the flowering season is recommended to increase pollination, promoting fruit set and production of more number of capsules.

Harvesting and Processing

Cardamom plants normally start bearing two years after planting. In most of the areas the peak period of harvest is during October-November. Picking is carried out at an interval of 15-25 days. Ripe capsules are harvested in order to get maximum green colour during curing which is indicated by dark green colour of rind and black coloured seed. Avoid harvesting of ripened capsules as it leads to the loss of green colour and also causes splitting of capsules during curing process. Immature capsules on processing yields uneven sized shriveled and undesirably coloured produce.

After harvest, capsules are dried either in fuel kiln or electrical drier or in the sun. It has been found that soaking the freshly harvested green cardamom capsules in 2 percent washing soda solution for 10 minutes prior to drying helps to retain the green colour during drying. When drier is used, it should be dried at 45 to 50°C for 14 to 18 hours, while for kiln, over night drying at 50 to 60°C is required. The capsules kept for drying are spread thinly and stirred frequently to ensure uniform drying. The dried capsules are rubbed with hands or coir mat or wire mesh and winnowed to remove any foreign matter. They are then sorted out according to size and colour, and stored in black polythene lined gunny bags to retain the green colour during storage. These bags are then kept in wooden chambers. Efficient and highly automated cardamom driers have been developed and being widely used with alternative sources of fuels such as kerosene, liquid petroleum gas (LPG) and diesel or with combination of fuels. Such kind of improved systems has the advantage of retaining high quality of produce with respect to colour and duration of curing is also substantially reduced by 16- 18 hours.

Dried capsules are polished either manually or with help of machines. Polishing is carried out by rubbing the dried capsules in hot state against a hard surface. The polished produce is subsequently graded based on the quality parameters such as colour, weight/volume, size and percentage of empties, malformed, shriveled and immature capsules.

Grades and specifications for Indian cardamom					
Grade	Description	Size (mm)	Weight (g/L)	Colour	General characteristics
AGB	Extra Bold	7	435	Green	Kiln dried, 3 cornered and with ribbed appearance
AGS	Superior	5	385		
AGS 1	Shipment	4	320-350	Light green	
AGL	Light	3.5	260		
CGEB	Extra Bold	8	450	Golden to light green	Round, ribbed or smooth skin
CGB	Bold	7.5	435		
CG-1	Superior	6.5	415	Light green	
CG-2	Mota, Green	6	385	Green	
CG-3	Shipment	5.5	350	Cream	
CG–4	Light	3.5	280	Brown	
BL–1	–	8.5	340	Pale	Fully developed round, 3 cornered ribbed or smooth skin
BL–2		7	340	Creamy	
BL–3		5	300	Dull white	
AG – Alleppey Green, CG – Coorg Green, BL – Bleached					

Plant protection: Important pests and diseases affecting cardamom are given below with their typical damages/symptoms and control measures.

Pest/Disease	Damage/symptoms	Control measures
Pests **Thrips** (*Sciothrips*)	Adults and nymphs suck sap and cause damage to leaves, shoots, inflorescence, thrips affected capsules fetch lower price.	Regulate shade in thickly shaded area, spray quinalphos or phenthoate @ 200 or 150 ml in 100 litres of water and repeat the spray at 30-40 days. Spraying during afternoon hours is recommended to avoid damage to bees.
Shoot, panicle, capsule/borer (*Conogethes punctiferalis*)	Larvae bore the unopened leaf buds, panicles causing drying or feed on young seeds causing the capsules empty, incidence occur throughout year, but more pronounced during March-April, May-June and Sept. -October.	
Parasitic nematodes (*Meloidogyne incognita*)	Occurs in nursery and main field. Poor germination and establishment in the nurseries, stunted and poor growth of plants, shedding of immature capsules in the main field, heavy galling and abnormal branching of roots.	Treat the plants in the nursery with carbofuran 3 g @ 1.5 kg a.i. /ha or in the main field with carbofuran @ 5 g a.i./clump and apply 0.5 kg of neem cake per clump twice a year.
Diseases Katte diseases (Viral disease)	Spindle shaped, slender chlorotic flecks appear on youngest leaves, later these develop into pale green discontinuous stripes as leaves mature, mosaic symptoms are marked. Infected clumps are stunted, smaller in size, with slender tillers and shorter panicles.	Use healthy seedlings, rougue the infected plants. Spraying insecticides like dimethoate @ 0.05% to control aphids (*Pentalonia nigronervosa*) and other sucking pests (vectors).
Azhukal or capsule rot (*Phytophthora meadii*) (*P. nicotianae var. nicotianae*)	Occurs in heavy rain zones, affected capsules turn brownish black in colour, often rotting extends to tillers and rhizomes also.	Do trashing, remove infected and dead plants etc. during premonsoon months, Spray 1% Bordeaux mixture during May and repeat again in August. Alternatively, fungicides like (fosetyl – aluminium) (0.2%) or Potassium phosphonate (0.5%) can be sprayed @ 500 – 750 ml per plant.

		Drenching plant basin with Copper oxychloride (0.2%) reduces the soil inoculum and further spread of the disease. *Trichoderma viride* or *Trichoderma harzianum* mass multiplied on suitable carrier media may be applied to plant basins @ 1 kg during May and September – October. If the soil is drenched with COC or other fungicides, *Trichoderma* should be applied only after 15 days.
Damping or rhizome rot (*Pythium vexans, Rhizoctonia solani and Fusarium*)	Excessive soil moisture and poor drainage favour this disease in the nursery, infected seedlings collapse at collar region and die in patches, and entire clump dies in grown up plants.	Pre treat the nursery with 1:50 formaldehyde, drench the soil after germination with 0.2% copper oxychloride. Application of antagonistic *Trichoderma spp.* As mentioned in case of azhukal disease management effectively controls the disease in plantations.

2.3. GINGER (*Zingiber officinale* L.)

Family: **Zingiberaceae**

Ginger, an indigenous plant, is an important spice crop of the world. It is valued in medicine as a carminative and stimulant of the gastro-intestinal tract. Dry ginger is used for the manufacture of oil, oleoresin, essence, soft drink, non- alcoholic beverages and vitaminised effervescent soft drinks.

In India, ginger is grown in an area of 1,58,800 ha with a production of 7,56,880 tonnes of dry ginger, sharing 40 % of world production. Kerala, Megalaya and Orissa account for more area under ginger though it is grown in almost all states in India. India is the largest producer and exporter to more than 50 countries accounting for more than 70 percent of world production. India is annually exporting around 15700 tonnes, earning a foreign exchange of about Rs.1213 crores per annum. China is our major competitor in the world market.

Botany

Ginger is a herbaceous perennial with underground rhizomes having serial leafy shoots of 0.5 to 0.75 m height; leaves sheathy, alternately arranged, linear with 15 cm long and sessile flowers borne on a spike, condensed, oblong and cylindrical with numerous scar bracts; flowers numerous yellow in colour with dark purplish spots, bisexual, epigynous, stamens only one, ovary inferior, three carpelled; fruit an oblong capsule, seeds glabrous and fairly large.

Climate and Soil

Ginger grows best in warm and humid climate. It is mainly cultivated in the tropics from sea level to an altitude of 1500m, both under rainfed and irrigated conditions. For successful cultivation of the crop, a moderate rainfall at the sowing time till the rhizomes sprout, fairly heavy and well-distributed showers during the growing period, and dry weather with a temperature of 28° to 35°C for about a month before harvesting are necessary. Prevalence of high humidity throughout the crop period is desirable. Ginger thrives best in well drained soils like sandy or clay loam, red loam or lateritic loam. A friable loam, rich in humus is ideal. However, being an exhaustive crop, soil should be rich in fertility.

Varieties

Several cultivars of ginger are grown in the different ginger growing areas in India. They are generally named after the localities or places where they are grown. Some of the more prominent indigenous types are Maran (Assam), Kuruppampadi, Ernad and Wynad local (all from Kerala). A high yielding introduction **Rio-de-Janeiro** has become very popular among the growers. Its yield potential is 25 to 35 tonnes per hectare. The fibre content is 5.19 percent and dry ginger recovery is 16-18 percent.

S. No.	Name of the Variety	Parentage	Fresh Yield (t/ha)	Duration in days	Special attributes
1.	Suprabha	A clonal selection from Kanduli local	16.60	229	Profuse tillering, plumpy fingers, bright grey skin, fibre 4.4%, oil 1.9%, oleoresin 8.9%.

2.	Suruchi	A clonal selection from Kunduli local	11.60	218	Rhizome greenish yellow, fibre 3.8%, oil 2.0%, oleoresins 10%.
3.	Suravi	An x-ray Induced mutant of local cultivar	17.50	225	Fingers cylindrical rhizomes dark grazy skin, fibre 4%, oil 2.1%, oleoresin 10.2%
4.	Himagiri	Clonal selection from Himachal collection	13.50	230	Best for green ginger, less susceptible to rhizome rot disease, suitable for rainfed condition. 4.29% oleoresin, 1.6% essentials oil, 6.05% crude fibre, 20.2% dry recovery.
5.	IISR Varada	Selection from germplasm	22.6	200	High yielder, high quality bold rhizome, low fibre content (3.29% to 4.50 %), essential oil 1.7%, 6.7% oleoresin and 19.5% dry recovery, tolerant to diseases.
6.	IISR Rejatha	Selection from germplasm	22.40	200	High yielder, plumpy and bold rhizome, 2.36% essential oil, 6.3% oleoresin, 4.0% crude fibre, 20.8% dry recovery.
7.	IISR Mahima	Selection from germplasm	23.20	200	Higher yielder, plumpy bold rhizomes, 1.72% essential oil, 4.5%

					oleoresin, 3.26% crude fibre, dry recovery 23.0%.
8.	Athira	Somo clonal variant of Maran	21. 00	–	High yield and quality , bold rhizome, less crude fibre (3.40 %),tolerant to soft rot and bacterial wilt
9.	Karthika		19.00		High yield and high pungency, suitable for extraction of oleoresins, medium bold rhizomes with high recovery of ovolatile oil and oleoresins. tolerant to soft rot and bacterial wilt

N.B. 1 to 3 released from High Altitude Research Station, Pottangi (Orissa) and 4 released from Dept. of vegetable crops, YSPUH & F. Nauni, Solan, 5 to 7 released from IISR, Calicut, 8 and 9 relesed from KAU, Thrissur

Season: The best time for planting ginger in West Coast of India is during the first fortnight of May with the receipt of pre monsoon showers, while in North Eastern states, it is during April. Under irrigated conditions, it can be planted well in advance during the middle of February or early March.

Preparation of Land and Planting

Preparation of land starts with the receipt of early summer showers. The land is to be ploughed 4 to 5 times or dug thoroughly to bring the soil to fine tilth. Weeds, stubbles, roots etc. are removed. Beds of about one metre width, 15 cm height and of any convenient length are prepared at an interspace of 40-50 cm in between beds. In the case of irrigated crops, ridges are formed 40 cm apart.

Ginger is always propagated by portions of the rhizomes, known as **seed rhizomes.** Carefully preserved seed rhizomes are cut into small pieces of 2.5 - 5.0 cm length weighing 20–25 g, each having one or two good buds. The seed rate varies from 1500 to 1800 kg per hectare from region to region. The seed rhizomes are treated with 0.3 percent mancozeb for 30 min, drained and planted at a spacing of 20-25 cm along the rows and 20-25 cm between the rows.

Manuring: At the time of planting, well decomposed cattle manure or compost at the rate of 25-30 tonnes per hectare along with 2 tonnes of neem cake is to be applied along with 50 kg P_2O_5 and 25 kg K_2O. They may be applied either by broadcast over the beds prior to planting or applied in pits at the time of planting. Besides, 75 kg of N/ha is recommended which is to be applied in two equal split doses at 40 and 90 days after planting. The plants are to be earthed up, after each top dressing with the fertilizers and beds rectified. In zinc deficient soils basal application of zinc fertilizer up to 6 kg zinc / ha (30 kg of zinc sulphate / ha) gives good yield.

Mulching: Mulching the beds with green leaves or organic waste is an important operation for ginger. Besides a source of organic manure, mulching prevents washing of soil, conserves soil moisture, smothers weed growth and improves the physical properties of the soil. The first mulching is done at the time of planting with 12.5 tonnes of green leaves and the second mulching is given after 46th day and 90th day with 5 tonnes of green leaves per hectare immediately after weeding and application of fertilizers. Daincha can be raised in the interspaces of beds immediately after planting ginger and they can be uprooted before second mulching and may be used for second mulching after earthing up.

After cultivation: Weeding is done just before fertilizer application and mulching. Two to three weeding are required depending on the intensity of weed growth. Proper drainage channels are to be provided when there is stagnation of water.

Rotation and mixed cropping: Being an exhaustive crop, it is not desirable to grow ginger in the same land continuosly for years and is hence commonly rotated with other crops. The crops most commonly rotated with ginger in Kerala are tapioca, chillies, dry paddy and gingelly in rainfed areas and ragi, groundnut, maize and vegetables in irrigated conditions. In Karnataka, ginger is also cultivated with ragi, redgram and castor. Ginger is also grown as an intercrop in coconut, arecanut, coffee, orange, mango, and guava plantations.

Harvesting and Curing

Harvesting is done from 6th month onwards for marketing the produce as green ginger. The rhizomes are thoroughly washed in water two or three times to remove the soil and dirt and sun dried for a day.

For prepring the dry ginger, the crop is harvested between 245 to 260 days. When the leaves turn yellow and start gradually drying up, the clumps are lifted carefully with a spade or digging fork and the adhering soil removed. The average yield per hectare varies from 15 to 25 tonnes.

For preparing dry ginger, the produce is kept soaked in water overnight. The rhizomes are then rubbed well to clean them. After cleaning, the rhizomes are removed from the water and the outer skin is removed with bamboo splinters having pointed ends. The peeled rhizomes are washed and dried in sun uniformly for one week. The dry rhizomes are rubbed together in order to get rid of the last bit of the skin or dirt. These are called **unbleached ginger.** To get good appearance, peeled rhizomes are soaked in 2 per cent lime water for 6 hours and then dried and this is known as **bleached ginger.** The yield of dry ginger is 16-25 percent of the fresh ginger depending on the variety, location etc.

Once ginger is dry, it is sorted and graded based on size of rhizome, its colour, shape, extraneous matter. presence of light pieces and extent of residual lime (in bleached ginger). Major grades are Garbled Non bleached Calicut (NGK), Ungarbled Non bleached Calicut (NUGK); Garbled Non bleached Cochin (NGC), Garbled Bleached Cochin (BGC), and Ungarbled Bleached Calicut (BUGK) etc. In International trade, Jamaican ginger is the first grade, followed by Cochin ginger.

Seed rhizomes: In order to get good germination, the seed rhizomes are to be stored properly in pits under shade. Large and healthy rhizomes from disease free plants are selected immediately after the harvest for seed. The seed rhizomes are treated with a solution containing 0.1 per cent quinalphos and 0.3 per cent Dithane M 45 for 30 minutes and dried under shade. The seed rhizomes are stored in pits of convenient size in sheds.

The walls of the pits may be coated with cowdung paste. The seed rhizomes are put in pits in alternate layers with dry sand or saw dust. Sufficient gap is to be left at the top of the pits for adequate aeration. The pits can be covered with wooden plank with one or two small holes for aeration. The seed rhziomes can also be stored in pits dug in the ground under shade. In some areas, the rhizomes are loosely heaped over a layer of sand or paddy husk, straw covered with dry leaves in a thatched shed.

Plant Protection

Pests/Diseases	Symptoms	Control measures
Shoot borer (*Conogethes punctiferalis*)	Larvae bore into the pseudostems and feed on growing shoots, yellowing and drying of the infested shoots.	Spray malathion 0.1% during July-October at monthly intervals.
Leaf roller (*Udaspes folus*)	Rolls the leaves and feeds on them.	Spray carbaryl 0.1%
Rhizome scale (*Aspidiella hartii*)	Infests rhizome, feeds on plant sap, cause withering.	Dip the seed rhizomes in quinalphos 0.1% twice prior to storage/sowing.
Diseases Soft rot or rhizome rot (*Pythium aphanidermatum*)	Collar region exhibits rotting and it spreads to rhizome and root. Leaves exhibit yellowing symptoms.	1. Provide good drainage and select healthy, disease free seed rhizomes. 2. Treat seed rhizomes with 0.3% Mancozeb for 30 min. before storage and planting. 3. Drench the beds with 0.3% Mancozeb 4. Add neem cake 2t/ha as basal dressing. Application of *Trichoderma harzianum* along with neem cake @ 1 kg/bed.
Bacterial Wilt (*Pseudomonas solanacearum*)	Affected pseudostem or rhizome shows a milky ooze on gentle pressing.	Treat seed rhizome with 200 ppm streptocycline for 30 mts.
Nematodes (*Meloidogyne, Radopholus, Pratylenchus*)	Stunting, chlorosis, poor tillering, necrosis are common aerial symptoms.	Treat the seed rhizomes with hot water (50^0C), sterilize the bed for 40 days, and incorporate *Pochonia* chlamydosporia, bio control agent @ 20g / bed at the time of sowing.

2.4 TURMERIC (*Curcuma longa* L.)

Family: Zingiberacea

Turmeric is the dried rhizome of **Curcuma longa,** a herbaceous plant, native to tropical South East Asia. The rhizome has 1.8 to 5.4 percent *curcumin,* the pigment and 2.5 to 7.2 percent of essential oil. It is used as an important condiment and as a dye with varied applications in drug and cosmetic industries.

In India, it is grown in an area of 1,93,690 ha producing annually about 9,69,700 tonnes. Although, India is leading in its production (86% of world output), the average productivity and quality are not satisfactory and these limit our export to about 10 to 15 percent of our production only. However, annually 700 crores worth of turmeric are exported. In India, Andhra Pradesh is the leading state followed by Maharashtra, Tamil Nadu, Orissa, Kerala and Bihar.

Botany

It is a herbaceous perennial with a thick underground rhizome giving rise to primary and secondary rhizomes called **fingers.** The leaves are broadly lanceolate with long leaf stalks. The flowers are borne on separate peduncle arising directly from the rhizome.

There are four important species of *Curcuma.* They are (a) *Curcuma longa,* the widely cultivated type (b) *C.aromatica,* the cochin turmeric or kasturi manjal (c) *C. angustifolia,* East Indian Arrow root having plenty of starch in its rhizome and (d) *C.amada,* mango ginger, which has the taste and flavour of raw mango.

Climate and Soil

Turmeric can be grown in diverse tropical conditions from sea level to 1500 m in the hills, at a temperature range of 20 to 35°C with a rainfall of 1500 to 2250 mm per annum. It is also grown as an irrigated crop. It is grown in different types of soils from light black, sandy loam and red soils to clay loams, but it thrives best in a well drained sandy loam rich in humus content with a pH range of 4.5 to 7.5.

Varieties

Number of varieties are recognized in the country and are known by the name of locality where they are cultivated. Recently, a number of improved varieties have been released from different states (Table-1).

Preparation of land

The land is prepared with the receipt of early monsoon. Soil is brought to a fine tilth by giving about four deep ploughings. Weeds, stubbles, roots etc.

are removed. Immediately after the receipt of pre-monsoon showers, beds of 1 to 1.5m width, 15 cm height and of convenient length are prepared with a spacing of 40 to 50 cm between beds. Planting is also done by forming ridges and furrows.

Planting: Kerala and other West Coast areas where the rainfall is sufficiently early, crop can be planted during April-May with the receipt of pre-monsoon showers. In Andhra Pradesh and Tamil Nadu, sowing is done during May-June or July-August. Since turmeric is a shade loving plant, castor or *Sesbania grandiflora* may be raised along the border lines in the field.

Seed materials: Whole or split mother rhizomes weighing 35 to 44 g are used for planting. Well developed healthy and disease free rhizomes are to be selected. Rhizomes are treated with 0.3 percent Mancozeb and 0.5 percent malathion for 30 minutes before storing. Two system of planting viz., flat beds and ridges and furrows methods are adopted in India. Small pits are made with a hand hoe in the beds in rows with a spacing of 25 × 25 cm × 30 cm and covered with soil or dry powdered cattle manure. The optimum spacing in furrows and ridges is about 45 to 60 cm between the rows and 25 cm between the plants. A seed rate of 2500 kg of rhizomes is required for one hectare.

As the seed rhizome is the costliest input in turmeric cultivation, rapid and economic method of turmeric propagation is being followed in certain pockets. In this method, fresh seed rhizomes are to be treated with 0.2% solution of carbendazim for 2 minutes and each rhizome piece with a single bud is kept one above the other up to five to six trays in one rack and covered under moist gunny bag at the top. On 6th day, wet gunny may be removed and the trays are placed in the nursery evenly on the floor. The sprouts appearing white emerge fast up to 7 to 8 days and unfurl into green leaves. These may be maintained in semi shaded area until 20th day (2-3 full grown green leaves stage). Two days prior to planting, watering may be stopped to harden the seedlings. This method ensures only low seed rate (750 kg/ha), high establishment (> 95%), faster growth and higher production than the conventional direct planting method.

Manures and manuring: Farm yard manure @ 10 t/ha is applied as basal dressing. The other NPK recommendation followed for Tamil Nadu (irrigated) and Kerala (rainfed) are given below.

Table 1 Improved varieties in turmeric

S.No	Variety	Parentage	Yield (t/ha)	Duration (days)	Curcumin content (%)	Special features
1	CO 1	X-ray induced mutant from Erode Local	5.85	285 (L)	3.2	Suitable for drought, saline soils rhizomes-big sized, bright orange, essential oil 3.2%.
2	CO (Tur)-2	Clonal selection from germplasm type	41.9*	270-280	4.2	Moderately resistance to leaf blotch, leaf spot,rhizome rot , field tolerant to thrips and scales etc, recovery is 20%
3	BSR- 1	X-ray induced mutant from Erode Local	6.00	285 (L)	4.2	Suitable for waterlogged condition, bright yellow rhizomes, oil content: 3.7%.
4	BSR - 2	Induced mutant from Erode Local	32.7*	245 (S)		A high yielding variety with bigger rhizomes, resistant to scale insects
5	Suguna	A clonal selection from Assam type	7.20	190 (S)	4.9	Thick and plumpy rhizomes, oil content 6%. oleoresin 13.5%, essentialoil 6.9% and dry recovery 20.4%, field tolerance to rhizome rot
6	Suvarna	A clonal selection from Singhat, Manipur	4.60	210 (M)	4.0	Rhizome medium in size, dark orange colour, field tolerant to pest and diseases. oil content 7%. oleoresin 13.5%, and dry recovery 20.0%
7	Sudharsana	A clonal selection from Singhat, Manipur	7.29	190 (S)	7.9	Thick and plumpy rhizomes, High yielding variety, field tolerant to rhizome rot. oleoresin 15.0%, essential oil 7.0% and dry recovery 20.6%.

Continued

#	Name	Selection				Description
8.	IISR Prabha	Selection from germplasm collection from Assam	37.47*	205 (S)	6.5	High yielding variety, oleoresin 15.0%, essential oil 6.5% and dry recovery 19.5%,
9.	IISR Prathiba	Selection from germplasm collection from Assam	39.21*	225	6.2	High quality line, 16.2% oleoresin, 6.2% essential oil, 18.5% dry recovery
10.	IISR Alleppy Supreme	Selection from germplasm collection from Assam	35.45*	.210	5.5	Shows tolerance to leaf blotch disease, 16.0% oleoresin, 19.0% dry recovery,
11.	IISR Kedaram	Selection from germplasm collection from Assam	34.5*	210	5.5	Tolerant to leaf blotch disease, Rhizomes contain 13.6% oleoresin, driage 18.9%.
12.	Roma	A clonal selection from Sunder	6.43	253 (M)	9.3 6.1%	Suitable for both rainfed and irrigated condition. Suitable for hilly areas and late season planting 4.2% oil, oleoresin 13.2% and dry recovery 31.0% less susceptible to diseases.
13.	Suroma	A clonal selection from T.Sunder	5.0	253 (M)	9.3 6.1%	4.4% oil, oleoresin 13.1%, dry recovery 26.0%, round and plumpy mother rhizomes, reddish brown skin and field tolerance to leaf blotch, leaf spot and rhizomes scale.
14.	Ranga	A clonal selection from RajpUri local	7.0	250 (M)	6.3	4.4% essential oil oleoresin 13.5%, dry recovery 24.8%. Bold and spindle shaped rhizomes, orange yellow colour. Suitable for late planting. Moderately resistant to leaf blotch and scales.

Continued

15	Rasmi	A clonal selection from Rajpuri local	7.8	240 (M)	6.4	Oil content 4.4% oleoresin 13.4%, dry recovery 23.0%, Mother rhizomes round shaped with plumpy primary fingers. Suitable for rainfed and irrigated condition, yearly and late sown season.
16.	Kanthi	Clonal selection from Mydukur variety of Andhra Pradesh	37.65*	240-270(M)	7.19%,	oleoresin 8.25%, essential oil 5.15%, dry recovery 20.15%
17.	Sobha	Clonal selection from local type	35.88*	240-270 (M)	7.39%,	Dry age 19.38%, oleoresin (9.63%), essential oil (4.29%).
18.	Sona	Clonal selection from local germplasm	4.02	240-270 (M)	7.12%,	Orange yellow rhizome, medium bold with no tertiary fingers present, Field tolerant to leaf blotch. essential oil 4.4%, oleoresin 10.25%, 18.88% dry recovery,
19.	Varna	Clonal selection from local germplasm	4.16	240-270 (M)	7.87%,	Bright orange yellow rhizome, Field tolerant to leaf blotch. essential oil 4.56%, oleoresin 10.8%, 19.05% dry recovery,
20.	Krishna	A clonal selection from Tekurpeta (A.P)	4.00	255 (M) 240	2.8	Long and plumpy rhizomes, oil content 2.0% olerosin 3.8 %, dry recovery 16.4% moderately resistant to rhizome fly, leaf spot diseases.
21.	Sugundham	A clonal selection from Kerala type	4.00	210 (M)	3.1	Reddish yellow rhizomes, stout and long fingers with short internodes oleoresin 11.0%, essential oil: 2.7%, dry recovery 23.3% moderately tolerant to pest and disease.

Continued

22.	Rajendra Sonia	A selection from Bihar type	4.5	225 (M)	8.4	5.0% essential oil, dry recovery 18.0% stout and plumpy rhizomes, deep orange
23.	Megha Turmeric	Selection from local tpes	20.0*	300 – 315	6.8%	Suitable for North East hill and North West Bengal. Bold rhizomes, dry recovery 16.37%,
24.	Pant Peethabh	Clonal selection from local types	20.0*		7.5%,	Long attractive fingers, essential oil 1.0%, dry recovery 18.5%, resistant to rhizome rot
25.	Suranjana	Clonal selection from North Bengal	(Potl. Yield. 29.0)*	235	5.7%	Suitable for open and shaded conditions, solo or intercrop, suitable for rainfed as well as high rainfall areas. Oleoresin 10.9%, essential oil 4.1%, dry recovery 21.2%, tolerant to leaf blotch and rhizome rot. Resistant to rhizome scales and moderately resistant to shoot borer.

Note: S=Short duration, M=Medium duration, L=Long duration
1-4 released by TNAU, 5-11 released by IISR, Calicut, 16- 19 released by KAU, Vellanikkara, 20 - released by Maharashtra Agrl. University, Maharashtra , 21- Gujarat Agrl. University, Jagudan, 22 - by Rajendra Agrl. University, Dholi, 23 Released by ICAR, shilling, 24 Released by GB Pant University, Pantnagar, 25- Uttarbengal Krishi Vishwa Vidyalaya, west Bengal.
* fresh yield

Manure	Tamil Nadu (kg/ha)	Kerala (kg/ha)
Neem cake	200-basal	–
N	25 kg each at basal, 30, 60, 90 and 120 days after planting i.e., 125 kg N.	30 kg N, 20 kg N and 10 kg N at 40 and 90 days after planting respectively.
P_2O_5	60 kg as basal	30 kg as basal
K_2O	60 kg as basal	60 kg - half as basal and half at 90 days after planting
$FeSO_4$	30 kg basal	–

Zinc @ 5 kg/ha may also be applied at the time of planting and organic manures like oil cakes can also be applied @ 2 t/ha. In such case, the dosage of FYM can also be reduced. Integrated application of coir compost @ 2.5 t/ha) combined with FYM, biofertilizer (*Azospirillum*) and half recommended dose of NPK is also recommended.

Beds are earthed up each time after top dressing.

Fertigation with recommended NPK (150:60:108 kg/ha) is applied throughout the cropping period once in three days to maximize yield. 75 % of the recommended dose of phosphorous is applied as basal dose. Water soluble fertilizers like 19:19:19, Mono ammonium phosphate (12:61:0), Multi K (13:0:45) and urea are used. The following fertigation schedule is recommended:

Crop Stage	Duration (in days)	Nutrients requirement (%)			Quantity applied (kg/ha)	
Planting to establishment stage	15	10	20	10	19:19:19 Multi K Urea	15.78 17.33 21.20
Vegetative stage	60	40	30	20	19:19:19 Multi K Urea	9.83 96.00 100.57
Rhizome initiation stage	60	30	30	30	19:19:19 Multi K Urea	4.91 71.28 76.29
Rhizome maturation stage	135	20	20	40	19:19:19 Multi K Urea	15.78 40.42 47.06
Total Duration	270	100	100	100		

Mulching: The crop is to be mulched immediately after planting with green leaves or banana psuedostem or sugarcane trash at the rate of 12 to 15 tonnes per hectare. It may be repeated for second time after 50 days with the same quantity of green leaves after weeding and application of fertilizers.

After cultivation and growing as intercrop: Weeding may be done thrice at 60,120 and 150 days after planting depending upon weed intensity. It can be grown as an intercrop in coconut and arecanut plantations. It can also be raised as a mixed crop with chillies, colocasia, onion, brinjal and cereals like maize, ragi etc. In some places, double inter cropping viz., Fenugreek + Onion in turmeric field is followed. Depending on the soil types, irrigated crops require 15 to 20 irrigations in heavy soils and 35 to 40 in light soils. Moisture stress affects the growth and development of the plant especially during the rhizome bulking stage.

Harvesting

Depending upon the variety, the crop becomes ready for harvest in seven to nine months. Usually it extends from January-March. Early varieties mature in 7 to 8 months, medium varieties in 8 to 9 months and late varieties after 9 months.

The land is ploughed and the rhizomes are gathered by hand picking or the clumps are carefully lifted with a spade. Harvested rhizomes are cleaned of mud and other extraneous matter adhering to them. The average yield per hectare is 20 to 25 tonnes of green turmeric.

Preservation of seed rhizomes: Rhizomes for seed purpose are generally stored after heaping under the shade of a tree or in well ventilated shed and covered with turmeric leaves. Sometimes the heap is plastered with earth mixed with cowdung. The seed rhizomes can also be stored in pits with saw dust. The pits can be covered with wooden planks with one or two holes for aeration.

Processing

It involves three steps viz curing, polishing and colouring.

Curing: Fingers are separated from mother rhizomes and are usually kept as seed material. The fresh turmeric is cured before marketing. Curing involves boiling of fresh rhizomes in water and drying in the sun.

In the traditional method, the cleaned rhizomes are boiled in copper or galvanised iron or earthern vessels, with water just enough to soak them. In certain places, cowdung slurry is used as boiling medium. From hygienic point of view, such rhizomes fetch poor market value. Boiling is stopped when froth comes out and white fumes appear giving out a typical odour. The boiling lasts for 45 to 60 minutes when the rhizomes are soft. Over-cooking spoils the colour of final product while under cooking renders the dried product brittle.

In the improved scientific method of curing the cleaned fingers (approximately 50 kg) are taken in a perforated trough of size 0.9×0.55×0.4 m, made of GI or MS sheet with extended parallel handle (Fig. 6). The perforated trough containing the fingers is then immersed in the pan. The alkaline solution (0.1% sodium carbonate or sodium bicarbonate) is poured into the trough so as to immerse the turmeric fingers. The whole mass is boiled till the fingers become soft. The cooked fingers are taken out of the pan by lifting the trough and draining the solution into the pan. Alkalinity of the boiling water helps in imparting orange yellow tinge to the core of turmeric.

The drained solution in the pan can also be used for boiling another lot of turmeric along with the fresh solution prepared for the purpose. The cooking of turmeric is to be done within two or three days after harvesting. The mother rhizomes and the fingers are generally cured separately.

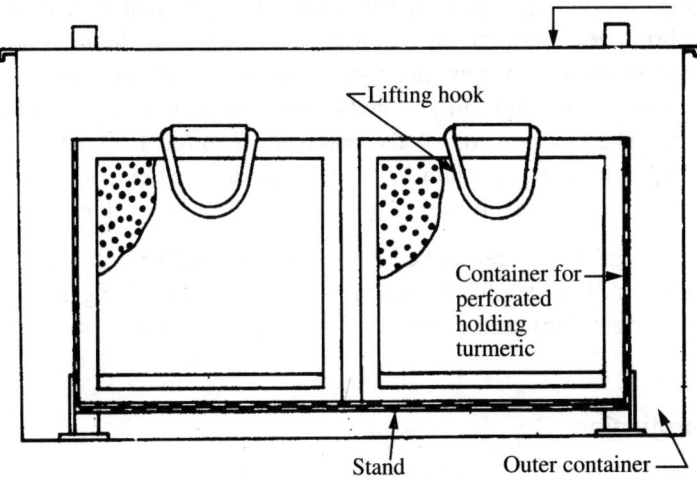

Figure 6 Turmeric boiling unit

The cooked fingers are dried in the sun by spreading 5 to 7 cm thick layers on bamboo mat or drying floor. A thinner layer is not desirable, as the colour of the dried product may be adversely affected. During night time, the materials should be heaped or covered. It may take 10 to 15 days for the rhizomes to become completely dry. The yield of the dry product varies from 20 to 30 percent depending upon the variety and the location where the crop is grown.

Polishing: Dried turmeric has poor appearance and a rough dull outer surface with scales and root bits. The appearance is improved by smoothening and polishing outer surface by manual or mechanical rubbing. Manual polishing consists of rubbing the dried turmeric fingers on a hard surface or trampling them under feet, wrapped in gunny bags. The improved method is by using hand operated barrel or drum mounted on a central axis, the sides of which are made of expanded metal mesh. When the drum filled with turmeric is rotated at 30 rpm, polishing is effected by abrasion of the surface against the mesh as well as by mutual rubbing against each other as they roll inside the drum. The turmeric is also polished in power-operated drums. The yield of polished turmeric from the raw materials varies from 15 to 25 percent.

Colouring: It is done to give a good appearance and better finish to the product. This is done to half polished rhizomes in two ways, known as dry and wet colouring. Turmeric powder is added to the polishing drum in the last 10 minutes in dry process. In wet process, turmeric powder is suspended in water and mixed by sprinkling inside the polishing basket. For giving a brighter colour, the boiled, dried and half polished fingers are taken in baskets which are shaken continuously when an emulsion is poured in. When the fingers are uniformly coated with the emulsion, they may be dried in the sun. The composition of the emulsion required for coating 100 kg of half boiled turmeric is Alum 0.04 kg, turmeric powder 2 kg, castor seed oil 0.14 kg, sodium bisulphate 30 g, concentrated hydrochloric acid 30ml.

Agmark grades are prescribed for whole turmeric and turmeric powder separately for export and for internal trade. Alleppy Finger Turmeric, Rajapore Finger Turmeric, Turmeric bulbs etc. are a few of the popular grades.

Plant protection

Pest/Disease	Symptoms	Control
Pest Shoot borer (*Conogethes punctiferalis*)	Larvae bore into the pseudostems and feed on the growing shoot resulting in yellowing drying of the infested shoots.	Spray malathion 0.1% at monthly interval from July to October.
Rhizome scale (*Aspidiotus hartii*)	Feed on plant sap in the field or on rhizome in storage, resulting in withering and drying.	Dip the rhizomes in quinalphos 0.1% twice prior to storage and sowing.

Disease		
Rhizome and root rot *(Pythium graminicolum)*	Complete drying of leaves, in advanced stages, rhizomes get decomposed and decayed.	Drench the soil with 0.1% wet cerasan.
Leaf blotch *(Taphrina maculans)*	Small oval, rectangular or irregular brown spots on either side of leaves.	Spray 0.2% Mancozeb.
Nematodes *(Meloidogyne sp., Radopholus similis)*	Leaves dry from tip, root system denied of feeder roots, crop remains stunted.	Apply to soil aldicarb or carbofuran granules at 1 kg a.i./ha.

CHAPTER

3

Seed Spices

3.1. CORIANDER (*Coriandrum sativum* L.)

Family: **Apiaceae**

Coriander (*Coriandrum sativum* L.) is an annual herb, mainly cultivated for its fruits as well as for the tender green leaves. It is native of the Mediterranean region and is now commercially grown in India, Morocco, U.S.S.R., Hungary, Poland, Rumania, Czechoslovakia, Guatemala, Mexico and the USA. In India, it is grown in Andhra Pradesh, Tamil Nadu, Karnataka, Rajasthan and Madhya Pradesh. It is grown in about 5,47,740 hectares with an annual production 5,27,530 tonnes of grains, producing nearly 87 % of world requirement. Though major portion is consumed locally, the export quantity is on the increasing trend every year.

The fruits have a fragrant odour and pleasant aromatic taste. The odour and taste are due to the essential oil content which varies from 0.1 to 1.0 percent in the dry seeds. These essential oils are used for flavouring liquors, cocoa preparations in confectionary and also to mask the offensive odours in pharmaceutical preparations. The dried ground fruits are the major ingredient of the curry powder. The whole fruits are also used to flavour foods like pickles, sauces and confectionary. The young plants as well as the leaves are used in the preparation of chutney and are also used as

seasonings in curries, soups, sauces and chutneys. It has medicinal properties too. Fruits are said to have carminative, diuretic, tonic, stomachic and
aphrodisiac properties.

Botany

Coriander belongs to the family Apiaceae (previously classified under family
umbelliferae). It is a smooth, erect annual herb 30 to 70 cm high, lower
leaves broad with crenately lobed margins, upper leaves finely cut with
lineary lobes, flowers small, white or pink in compound terminal umbels,
fruits - schizocarp, globular, yellow-brown, ribbed, 2 seeds, ripe seeds are
aromatic.

Climate and Soil

It is a tropical crop and can be grown throughout the year (except very
hot season i.e., March-May) for leaf purpose, but for higher grain yield
it has to be grown in specific season. A dry and cold weather free from
frost especially during flowering and fruit setting stage, favours good grain
production. Cloudy weather during flowering and fruiting stage favours pest
and disease incidences. Heavy rain affects the crop. As an irrigated crop,
it can be cultivated on almost all types of soils provided sufficient organic
matter is applied. Black cotton soils with high retentivity of moisture are
best under rainfed conditions.

Varieties

Improved varieties of coriander are now available for cultivation in Tamil
Nadu, Andhra Pradesh, Gujarat and Rajasthan states. They are described
briefly in Table 3.1

Table 3.1   Improved Varieties in Coriander

S.No.	Variety	Parentage	Special Characters	Duration (days)	Yield/ ha (kg)
1.	Col	A pure line selection from Koilpatti local	A variety with small statuted plants, Tall plant, many umbels per plant, suitable for rain fed areas and for green and grains, essential oil 0.27%	110	500

2.	CO 2	A reselection from culture P_2 of Gujarat	A dual purpose variety, suitable for saline, alkaline soil, High yield, dual purpose variety, tolerant to drought, seeds oblong medium, essential oil 0.27 %oil 0.3%	90-110	600-700
3.	CO 3	Reselection • from Acc.No. 695 from IARI, New Delhi	High yield, dual purpose variety suitable for saline, alkaline, soil suitable for rainfed and irrigated conditions, rabi as well as Kharif season. Field tolerant to powdery mildew, wilt and grain mould. Medium size grain, seed oil 0.38-0.41%	85-95	640
4.	Co (CR)4	Reselection from germplasm ATP 11 from Guntur collection	Early maturing variety, suitable for both rainfed and irrigated condition; grains oblong and medium; essential oil content is 0.4%; field tolerant to wilt and grain mould.	65-70	600
5.	Gujarat Coriander-1	A selection from local germplasm	Suitable for early sowing, erect plant, moderately tolerant to wild and powdery mildew. High yield, more number of branches, seeds bolder and greenish in colour, essential oil 0.35%	112	1100
6.	Gujarat Coriander-2	A selection from CO 2	Semi spreading type, suitable for early sowing, moderately tolerant to powdery	110-115	1500

			mildew, grains oblong, essential oil 0.40%, lodging and shattering resistant. High yield, more branches, dense, foliage, umbels large size, grain purpose variety, bold seeds, no lodging		
7.	Rajendra Swati	A mass selection from germplasm type (Mazaffarpur)	High yield potential, suitable for intercropping, fine seeded, rich in essential oil, resistant to stem gall disease.	110	1200-1400
8.	Rcr-41	Recurrent selection from UD41 (Kota)	High yield, tall erect, suitable for irrigated areas, resistant to stem gall, wilt and moderately resistant to powdery mildew; small, essential oil 0.25%	130-140	1200
9.	RCr 20	Recurrent half sib selection from Jaipur local	Medium sized bushly plant, suitable for rain fed crop or limited moisture conditions and heavy soils of south Rajasthan. Moderately resistant to stem gall, bold grains, essential oil 0.25%. Early maturity.	100-110	900
10.	RCr 435	Half sib selection from local germplasm from Jalore	Plants are bushy, erect, bold seeds, medium size, contain essential oil 0.33%; medium maturity; adapted for irrigated condition,	110-130	1000

			moderately resistant to root knot and powdery mildew. Medium maturity		
11.	RCr 436	Half sib selection germplasm from 'Kota'	Plants semi dwarf, bushy type; with quick early growth and bold seeds, resistant to root rot & root knot nematodes; most suitable for limited moisture condition and heavy soils of South Rajasthan, essential oil 0.33%	90-100	1100
12.	RCr 684	Mutation breeding by gamma rays. Induced mutant of Rcr-20	A variety, resistant to stem gall less susceptible to powdery mildew. Adapted to medium heavy textured soil, essential oil 0.32% and sandy loam soil under irrigations. Seeds of the variety are bold. Plants are tall and erect with higher number of seeds/umbel. Medium maturity	110-120	990
13.	RCr 446	Recurrent half sib selection from Jaipur local	Plants tall, are leafy erect with higher number of seeds per umbel. Seeds are medium in size and 0.33% volatile oil content. Moderately resistant to stem gall and wilt under rainfed, limited moisture or irrigation conditions, medium maturity	110-130	1200

14.	Swathi	Mass selection	High yield, medium size oval grain, semi erect, suitable for delayed sowing and rainfed condition, essential oil 0.30%, field resistant to diseases and white fly. It suits well to the areas where the soil moisture retentiveness is comparably less, being early maturity crop, it escapes powdery mildew disease.	80-90	885
15.	Sadhana	Mass selection from local Alur collection	High yield, suitable for rainfed areas, semi erect, field resistant to diseases and white fly, aphid and mites. Withstands moisture stress, responds well to input. Dual purpose variety.	95-105	1025
16.	Sindhu	Mass selection from germplasm Warangal local	Oval medium breakable grains, essential oil (0.4%); suitable for rainfed areas, tolerant to wilt, powdery mildew as well as drought condition.	100-110	1000
17.	Hisar Anand	Mass selection from Haryana Collection	A medium tall, dual purpose variety, oval medium size seeds; essential oil 0.35%, wider adaptability to different soil conditions. Resistant to lodging due to spreading habit.		1400

18.	Hisar Sugandh	Mass selection from indigenous germplasm	Suitable for irrigated conditions, Resistant to stem gall disease.		1400
19.	Hisar Surabhi	Mass selection from local germplasm	Bushy erect plant type, seed medium, oblong, oil content 0.425%, tolerant to frost, less susceptible to aphids.	130-140	1800
20.	Azad Dhania -1	Mass selection from Kalyanpur germplasm collection	Erect, early branching, number of umbellate per umbel is essential oil 0.29%, tolerant to moisture stress, tolerant to powdery mildew and aphids.	120-125	1000
21.	Pant Haritima	Selection from local type Pant Dhania	A tall erect plant, a dual purpose type, good yielder of leaf, smaller seeds with high oil (0.45%), resistant to stem gall.	155-160	1200
22.	DWA-3	Pure line selection from Karnataka collection	A dual purpose variety and for seed production in rabi crop, essential oil 0.27%, moderately tolerant to powdery mildew, black clay soils are best suited.		400
23.	Lam Sel. CS 2	Mass selection (Lam Sel CS 2)	Medium tall, bush type, more number of branches, grain purpose variety, tolerant to pests and diseases, good quality grain, 0.4% oil,	110	1500 kg/ha

24.	CIMPOS-33	Selection from germplasm introduced from Bulgaria	Tall erect, compact profusely branching and flowering, grains small and bold, essential oil 1.3%. Mainly recommended for oil production		2100
25.	NRCSS-ACR-1	Reselection from EC 467683 from Russia	Dual purpose variety, oil 0.35 – 0.5% long duration, resistant to stem gall and wilt.		1100

1 to 4 - Released from TNAU, Coimbatore; **5 and 6**- from GAU, Jagudan; **7** - from RAU, Dholi; **8 to 13** - from RAU, Jobner; **14 to 16, 23**- Released from Regional Agriculture Research Station, Guntur , **17-19**- released from CCSHAU, Hisar, **20**- from C.S. Azad Univ. of Agriculture and Technology, Kanpur, **21**- from GB Pant Univ., Pantnagar, **22**- from UAS, Dharwad; **24**- CIMAP, Bangalore; **25**- NRCSS, Ajmer.

Field Preparation

For raising a rainfed crop, the land is ploughed 3 to 4 times following rains and field must be planted immediately to break the clods and to avoid soil moisture. For irrigated crop the land is ploughed twice or thrice and beds and channels are formed.

Season of cultivation: In the North and Central parts of India and Andhra Pradesh, it is mostly grown as a Rabi season crop and hence sowing is done between middle of October and middle of November. Still late sowing is recommended in places vulnerable for frost damage. In certain pockets of the above area, late kharif crop is sometimes sown in August-September. In Tamil Nadu, as an irrigated crop, coriander is raised in June-July and September-October. In the first season, it matures early before the end of August-September. In the second season, the crop matures late with an extended growth phase during January- February. The growth and the yield of second season crop are found to be better than the first season crop. Under rainfed conditions, it is sown during September-October, at the onset of North-east monsoon and harvested during January-February.

Sowing: A seed rate of 10 to 15 kg per hectare is required. Seeds stored for 15 to 30 days record better and early germination than freshly harvested seeds. Seeds soaked in water for 12 to 24 hours before sowing also enhance better germination. The seeds are split into two halves by rubbing and treated with thiram at 2 g per kg of seeds. For irrigated crop, sowing is generally done in rows spaced at 30 to 40 cm apart with 15 cm between hills. Soil depth should not exceed 3.0 cm. Three to five seeds are sown in a hill and later on thinned to two plants per hill. Under rainfed conditions, seeds are broadcast and covered with country plough. Germination takes place in 10 to 15 days.

Manuring: About 10 tonnes of farm yard manure is applied at the time of last preparation. In addition, the following fertilizers may be applied.

Name	N		P	K
	Basal	**30 days after sowing**	**Basal**	**Basal**
Irrigated crop	15	15	40	20
Rainfed crop	20	–	30	20

Recently, integrated nutrient management practices involving FYM 5 t/ha + 50% recommended N through inorganic source + *Azospirillum* 10.5 kg/ha (as seed treatment) improves growth, vigour and seed yield. Similarly, micronutrient spray (0.5% $ZnSO_4$ and $FeSO_4$) at 45 and 60 days of sowing is also recommended to get higher yield.

Irrigation: First irrigation is given 3 days after sowing and thereafter at 10 to 15 days interval depending upon the soil moisture available in the soil.

After cultivation: The first hoeing and weeding are given in about 30 days. Thinning the plants is also attended simultaneously, leaving only two plants per hill. Depending upon the growth one or two more weeding are done.

Harvesting

The crop will be ready for harvest in about 90 to 110 days depending upon the varieties and growing season. In certain varieties, harvesting 50 percent leaves at 60 days and 75 days may be done which will fetch additional income but without affecting the grain yield. Harvesting has to be done when the fruits are fully ripe and start changing from green to brown colour. Delaying of the harvest should be avoided lest shattering during harvest and splitting of the fruits in subsequent processing operations. The plants are cut or pulled and piled into small stacks in the field to wither for 2 to 3 days. The fruits are then threshed out from the plants by beating with sticks or rubbing with hands. The produce is winnowed, cleaned and dried in partial shade. After drying, the produce is stored in gunny bags lined with paper.

The rainfed crop yields on an average 400 to 500 kg/ha and the irrigated crop 600 to 1200 kg/ha.

Off Season Production

During summer months, coriander leaves are in great demand (fetching very high price) as its production is limited due to the prevailing hot temperature (> 35°C). TNAU has developed a technology in which coriander is sown during March–April in shade nets (50 % shade) which provides as high as 100–120 kg/200 m^2 of leaves as against only 40–50 kg/200 m^2 under open field.

Plant Protection

At the seedling stage coriander is often attacked by the leaf eating caterpillars and semi-loopers and at the flowering stage by the aphids. Spraying the crop with chlorantraniliprol 18.5 % (0.25 ml/L) is recommended to control the aphids but at flowering stage the use of any insecticide would kill the bee population affecting pollination in the crop.

Powdery mildew *(Erysiphe polygoni)* is a serious disease which ruin the crop if allowed unchecked in the initial stage itself. Spraying wettable sulphur 0.25% or 0.2% solution of Karathane twice at 10 to 15 days interval is recommended. Grain mould is caused by *Helminthosporium* sp, *Alternaria* sp., *Carvularia* sp *and Fusarium* sp. It can be controlled by spraying carbendazim 0.1% 20 days after grain set.

3.2. FENUGREEK (*Trigonella foenum-graecum* L.)

Family: **Fabaceae**

Fenugreek, a native of South Eastern Europe and West Asia, is cultivated as a leafy vegetable, condiment and as medicinal plant. The fresh tender leaves and stem are consumed as curried vegetable and the seeds are mainly used as spice for flavouring almost all dishes. It has a high medicinal value as it prevents constipation, removes indigestion, stimulates spleen and liver and is apetizing and diuretic.

In India, it is grown in about 99,090 ha producing annually about 1,12,850 tonnes of seeds. We are exporting to Saudi Arabia, Japan, Sri Lanka, Korea and U.K., thus earning a foreign exchange worth of Rs.65 crores annually. The major states growing fenugreek in India are Rajasthan, Madhya Pradesh, Gujarat, Uttar Pradesh, Maharashtra and Punjab.

Botany

It is an annual herb reaching a height of about 0.9 m; leaves are light green, pinnately trifoliate, flowers - papilionaceous, fruits - legume, long, narrow, curved, tapering with a slender point and containing small deeply furrowed seeds. There are two species of the genus *Trigonella* which are of economic importance viz., *T. foenum graecum,* the common methi and *T. corniculata,* the Kasuri methi. These two differ in their growth habit and yield. The latter one is a slow growing type and remains in rosette condition during most of its vegetative growth period.

Varieties

Many improved cultivars are now available for cultivation. They are briefly described hereunder.

S.No	Variety	Parentage	Special Characters	Crop duration	Grain yield (kg per ha)
1.	CO 1	Reselection from TG 2336	Dual purpose quick growing, suited for intercropping, high seed protein 20-23%	90	685 kg grain, 4000 kg of green
2.	CO2	Selection from CF 390	Short duration dual purpose variety, field tolerant to Rhizoctonia root rot disease, suitable for both kharif and rabi season. Early maturity.	85-90	480
3.	Rajendra Kanti	Pureline selection from Reghunathpur collection	High yield, medium height, bushy, suited for pure as well as intercropping. Field tolerant to cercospora leaf spot, powdery mildew and aphids. Seed Protein: 9.5%.	120	1200-1400
4.	Hisar Sonali	Pure line selection from local germplasm	Tall and bushy vigorous growing, variety, dual purpose variety, late maturity suitable for	140-150	1700

			cultivation under irrigated condition, moderately resistant to leaf spot disease and root complex diseases.		
5.	Hisar Suvarna	Pure line selection from local germplasm	A quick growing, erect and tall, dual purpose, late maturity, moderately resistant to root rot and aphids.	130-140	1600
6.	Hisar Madhavi	Pureline selection from local germplasm of UP	A quick growing, erect and tall, dual purpose, medium maturity, moderately resistant to downy mildew and powdery mildew.	130-140	1900
7.	Hisar Mukta	Pureline selection Natural green seed coated mutant line from UP	A quick growing, seed type variety, medium maturity, moderately resistant to downy mildew and powdery mildew. Suitable for both irrigated and rainfed conditions.	135-140	2000
8.	Gujarat Methi I	Recurrent selection based on pureline selection from J.Fenu 102.	Dwarf variety		1860
9.	RMt –I	Pure line Selection from Nagpur Local	High yield, moderately branched, moderately tolerant, to root rot and powdery mildew. Seed Protein: 21%.	145	1500
10.	RMt 143	Pureline selection of local collection of Jodhpur	Moderately resistant to powdery mildew. Seeds bold yellow colour, suitable for heavier soils.	140-150	1600

11.	RMt 303	Mutation breeding from RMt1	Medium maturity, seeds bold less susceptible to powdery mildew.	145-150	1900
12.	RMt 305	Mutation breeding from RMt1	Determinate type multi poded, early maturing, resistant to powdery mildew, and root knot nematode	120-125	1300
13.	Lam Sel.1	A selection from Germplasm collection from UP	Dual purpose, High yield, bushy plant type. Seed Protein: 53%.	85-90	740
14.	Pant Ragini	Selection from local germplasm	Dual purpose, tall bushy type, resistant to downy mildew and root rots.	170-175	1200
15.	AM01 35 NRCSS AMI	Selection from local germplasm	Dual purpose, tolerant to powdery mildew.	-	1720

1 and 2- released from HC&RI, TNAU, Coimbatore; **3 and 4-** released from College of Agri., Dholi, Bihar; **5 to 7-** released from CCS, HAU, Hisar; **8-** released from GAU, Jagudan; **9 to 12-** released from SKN College of Agri., Jobner, Rajasthan; **13-** released from Regional Agri. Research Station ANGRAU, Guntur; **14-** released from GB Pant Univ. , Pantnagar; **15-** released from NRC Seed Spices, Ajmer

Climate and Soil

It has a wide adaptability and is successfully cultivated both in the tropics as well as temperate regions. It is tolerant to frost and freezing weather. It does well in places receiving moderate or low rainfall areas but not in heavy rainfall area. It can be grown on a wide variety of soil but clayey loam is relatively better. The optimum soil pH should be 6.0 to 7.0 for its better growth and development.

Land Preparation and Sowing

Land is prepared by ploughing thrice and beds of uniform size are prepared. Broadcasting the seed in the bed and raking the surface to cover the seeds is normally followed. But, line sowing is advocated in rows at 20 to 25 cm apart which facilitates the inter cultural operations.

Sowing in the plains is generally taken up in September to November while in the hills, it is grown from March. Approximately 20 to 25 Kg of seed is required for one hectare and the seed takes about 6-8 days to complete its germination.

Manures and fertilizers: Besides 15 tonnes of farm yard manure, a fertilizer dose of 25 Kg N, 25 Kg P_2O5 and 50 Kg K_2O per ha is recommended. Half of the N dose, and the entire quantity of P and K are applied basally and the remaining half N is applied 30 days after sowing. To obtain more successful leafy growth, nitrogen should be applied after each cutting.

Irrigation: First irrigation is given immediately after sowing and subsequent irrigation is applied at 7 to 10 days interval.

Intercultivation: Hoeing and weeding during the early stages of plant growth are required to encourage proper growth. Thinning may be done on 20 to 30 days to keep the distance between the plants at 10 to 15 cm and to retain 1 to 2 plants per hill.

In about 25 to 30 days, young shoots are nipped off 4 to 5 cm above ground level and subsequent cuttings of leaves may be taken after 15 days. It is advisable to take 1 to 2 cuttings before the crop is allowed for flowering and fruiting. When pods are dried, the plants are pulled out, dried in the sun and seeds are threshed by beating with stick or by rubbing with hands. Seeds are winnowed, cleaned and dried in the sun. They may be stored in gunny bags lined with paper.

Yield

A yield of 1200 to 1500 kg of seeds and about 800 to 100 kg of leaves may be obtained per hectare in crops grown for both the purposes.

Plant protection

Root rot (*Rhizoctonia solani*) is a serious disease and can be controlled by drenching carbendazim 0.05% first at the onset of the disease and another after one month.

3.3. CUMIN (*Cuminum cyminum* L.)

Family : **Apiaceae**

Cumin or "Safaid zeera", one of the oldest spices known since bibilical times, comprises the dried pale-yellowish seeds of a slender annual herb, believed to be a native of Egypt, Syria and Turkestan. Today the plant is grown extensively in Iran, India, Morocco, China, Southern Russia, Southern Europe and Turkey. According to recent statistics, India ranks first in the world in area and production of cumin. It is grown in about 5,93,980 ha in India with an annual production of about 3,94,330 tonnes, accounting for nearly 85 % of world production. India is also the largest consumer of cumin and exports hardly four per cent of its total production. Major types of cumin in the international trade are Iranian cumin, Indian cumin, Egyptian cumin and Turkish cumin. In India, cumin is extensively grown in Gujarat, Rajasthan, Madhya Pradesh, Uttar Pradesh, Punjab and Maharashtra.

Seeds yield a volatile oil (2.5 to 4.5 %), the chief constituent of which is cuminaldehyde (20-40%) which is used in perfumery. In addition to this, seeds also contain a fixed oil (10%) with a strong aromatic flavor. Hence, they are largely used as a condiment and form an essential ingredient in all mixed spices and curry powders. They also act as a stimulant, carminative, stomachic and astringent. They are now widely used in veterinary medicine also. The volatile oil is used in perfumery and also for flavouring liquors and cordials. Fixed oil finds use in fat and soap industry.

Botany

Cuminum cyminum is a slender much branched annual herb about 25cm in height, leaves 2-3 partite linear, deep green, flowers white or pink in colour produced in small compound umbels. The aromatic seed like fruit, commonly known as seed is elongated, oval, elliptical, deeply furrowed, approximately 6mm long, light yellowish brown. The odour is peculiar, strong and heavy, while flavour is warm, slightly bitter and somewhat disagreeable.   .

Varieties

Four group of cultivars viz. (1) tall, (2) dwarf, (3) pink flowered and (4) white flowered are generally found. The pink flowered variety yields better than the white flowered variety. Crop improvement work led to the release of following improved cultivars:

S.No	Variety	Parentage	Special characters	Yield/ ha (kg)
1.	S-404	Selection from local germplasm	An erect plant, medium size fruit, contain 2.2% essential oil, 7.7% crude fibre, moderately tolerant to powdery mildew.	350

2.	Mc.43	Selection from germplasm	Plants semi spreading, grains bolu lustrous, withstand lodging and shattering, moderately tolerant to Fusarium wilt, Alternaria blight and powdery mildew. Essential oil content 2.7%, crude fibre 15.5%.	580
3.	Gujarat Cumin 1	Selection from local germplasm (Vijaypur-5)	Plants bushy and spreading, grains bold, linear oblong, withstand shattering and lodging. Moderately tolerant to wilt, blight and powdery mildew. Essential oil 3.6%, crude fibre 14.25%.	550
4.	Gujarat Cumin 2	Pureline selection from M2 irradiated seeds of MC 43	Plants bushy with good branching habit. Grains bold, lustrous, medium sized. Tolerant to wilt, blight and powdery mildew. Essential oil content 4%, crude fibre 22.1%.	620
5.	Gujarat Cumin 3	Recurrent selection derived from W. German entry EC 232689	Bushy dwarf plant, fruit medium sized, frost and wilt resistant variety suitable for winter season. Essential oil content 4.4%.	620
6.	Gujarat Cumin 4	Natural crossing followed by selection	The first wilt resistant variety, normal seed appearance and no seed splitting habit.	1250
7.	RZ-19	Recurrent single plant progeny from collection of Ajmer	Erect plants, pink flowers, bold, lustrous grain, and gray pubescent, tolerant to wilt and blight. Suitable for late sowing.	560
8.	RZ-209	Recurrent single plant progeny from collection of Jalore	Moderately resistant to blight and wilt.	650
9.	RZ-223	Mutation breeding in UC-216	Wider adaptability, resistant to wilt, superior in yield and quality over RZ-19. Plants semi erect, long, bold, attractive seeds. Essential oil 3 to 3.5%.	600
10.	AC-01-167-2005	Released from EC-243373	Bold seeds, resistant to wilt, essential oil 3%	515

N.B 1-6 released from GAU, Jagudan, 7 to9 released from RAU, Jobner, 10 released from NRC seed spices, Ajmer

Soil and Climate

Cumin thrives well in tropical and subtropical climate. Frost is the most dangerous and limiting factor for cumin cultivation. High humidity during flowering and fruiting induces development of diseases like powdery mildew and blight. Sandy loam or loamy soil is supposed to be the best for its successful production.

Seeds and Sowing

The land is ploughed to fine tilth, leveled and the field is divided into beds of convenient size. Seed rate is 90 kg/ha. In Gujarat and Rajasthan, sowing is taken up between 15 November and late December. Late sowing results in poor yield. Seeds are broad cast uniformly and covered with soil lightly. In order to avoid fungal diseases, seed treatment with fungicides is done. Soaking of seeds for 24-36 hours in running water improves the germination percentage. Light irrigation is given after sowing. Line sowing can also be done at a spacing of 30 x 15 cm. It takes about a week to germinate.

After Cultivation

Well rotten FYM is applied at the rate of 15-20 t/ha basally. NPK at the rate of 70:60:40 kg/ha is applied basally and 30 kg /ha of nitrogen is top dressed 30-40 days after sowing.

First irrigation is given just after sowing, second 7-8 days after sowing. After 2-3 irrigations, the crop is irrigated at an interval of 12 to 20 days depending upon the weather conditions and soil type. Irrigation at the flowering and seeding stages are essential, however, it is to withhold at the time of seed maturity.

Thinning is done when the seedlings attain a height of about 5 cm (usually one month) maintaining a distance of 10 cm between plant to plant. First hoeing and weeding is done along with this.

Harvesting and Yield

The crop matures 100-110 days after sowing and leaves turn yellow. Plants are uprooted and stacked in small bundles for sun drying. The grains are separated by beating with light sticks and cleaned by winnowing. Yield is about 550 kg/ha. The clean and dried seeds are stored in gunny bags.

Processed products include cumin powder, cumin oleoresin, volatile oil and fixed oil etc.

3.4. MINOR SEED SPICES

Common name, Botanical name and family	Varieties	Cultural hints	Harvesting and yield	Other aspects
Fennel (*Foeniculum vulgare* Mill.) Apiaceae	Col, Gujarat Fennel-1, PF-S-7-9, PF-35, Gujarat Fennel-2, RF 101, RF 125, Hisar Sawrup, Azad Sanuf 1, Pant Madurika, Rajendra Sourabha, AF 01 119 NRCSS-AF 1, RF 143	Well drained loamy or black soil, mild climate, sea level to 2500 m MSL, direct sowing (9-12 kg of seeds/ha) or transplanting (3-4 kg/ of seeds raised in 100 m² nursery) 5 to 6 weeks old seedlings, spacing 60x30 cm, season-October/November in plains and May-June in hills, FYM10 t/ha N:90 kg in 3 splits (basal,30 and 60 days), 10 kg P and K basal, 2 or 3 weeding, thinning 20 days for sowing.	Crop matures in 7 to 8 months, harvest before fruits are fully ripe i.e., fruits turn to yellow colour, cut the stems along with umbel and dry in sun for 4-5 days, thresh the fruits and clean by winnowing, yield 250-400 kg/ha.	An aromatic herb, leaves used as sauce, dried fruit fragrant odour and pleasant aromatic taste, used for flavouring many dishes, fruits contain volatile oil (0.7 to 1.2%).. also used as a flavouring agent in many culinary preparations.
Celery (*Apium graveolens L.*) Apiaceae	RRL-85-1	Fertile light soil, mild cool climate, seed rate 375-500 g/ha, spacing 15x40 cm, 40-60 days old seedlings are plated, manuring 200 kg N and 30 kg P/ha, 2 or 3 light irrigations.	Pick the matured umbels periodically and collect the seed, yield 175-275 kg/ha	Mostly cultivated as a salad crop, the dried fruits are used as spice; Fruits yield 2-3% of volatile oil, used as a fixative and ingredient of novel perfumes. Seeds have much medicinal value.

Continued ....

72

Crop	Variety	Soil, climate, cultivation	Harvest and yield	Uses
Anise or Aniseed (*Pimpinella anisum L.*) Apiaceae	AN01-2	Fertile light soil, cool climate free from frost, seed rate 5-10 kg/ha, spacing 60×15 cm.	Duration 100-110 days, harvest when tips of fruit turn greyish green, and fruiting umbels are cut off, tied in bundle and stacked. After 4-5 days, fruits ripen; thresh out, yield 450-650 kg fruits/ha.	An annual herb, the seed has a sweet aromatic taste and is used for flavouring, confectionaries, baking materials etc., essential oil (2.7%), called 'oil of anise'.
Mustard (*Brassica nigra L. Koch*) Brassicacea		Well drained soil, mild cool climate, seed rate 6-7 kg/ha, spacing 45×30 cm in beds, basal FYM 10 t/ha. N 50 kg (basal and top) and P 60 kg (basal alone), 2 to 3 weeding.	90-120 days for maturity, when pods turn brown plants are pulled, dried in the sun threshed, yield 500 to 600 kg/ha.	Used as a condiment in South India and also used in pickles and curries. The seeds yield 27-33% of fixed oil, which is exclusively used as cooking oil in North India.
Dill (*Anethum graveolens. L.*) Apiaceae	Gujarat Dill-1, Guj. Dill-2, RSP-11, Ajmer Dill-1, Ajmer Dill-11.	Well-drained soil, mild climate, seed rate 3-4 kg/ha, sowing by October / November, manuring 10 cart load of FYM as basal, germination within 10-12 days, thinning in 3 weeks time.	Duration 130-150 days, plants are pulled when they become pale yellow, dried in sun and threshed, yield 800-900 kg/ha.	Herbaceous annual, seed used as a condiment, foliage flavouring agent, seed contains 1.5-4% of essential oil, used in soap industry.

CHAPTER

4

Tree Spices

4.1. CLOVE (*Syzygium aromaticum* L.)

Family: **Myrtaceae**

Clove, the dried unopened flower buds of the evergreen tree, *Syzygium aromaticum* (Syn. *Eugenia caryophyllus*) is an important spice noted for its flavour and medicinal values. It is indigenous to Moluccas Island (Indonesia) and was introduced to India around 1800 A.D. by the East India Company in their spice garden in Courtallam, Tamil Nadu. The major producers of this spice today are Indonesia, Zanzibar and Madagascar. World production is estimated to be 63,700 tonnes. Indonesia alone accounts for 66 per cent of the world production.

In India, it is grown in about 2134 ha. producing annually about 922 tonnes, however to meet our requirement we are importing about 10,000 tonnes worth of Rs. 45,000 lakhs. The important clove growing regions in India now are the Nilgiris, Tirunelveli and Kanyakumari districts of Tamil Nadu(765ha), Calicut, Kottayam, Quilon and Trivandrum districts of Kerala (1123ha)and South Kanara district of Karnataka (90ha) besides Andaman has about 196 ha.

Food processing industry uses both whole and ground form of cloves in various preparations. Clove oil is used in perfumeries, pharmaceuticals and flavouring industries. Clove oleoresin is also increasingly used in the food processing industry. In Indonesia, the major part is absorbed for making KRETEK cigarette industry.

Botany

The clove is an evergreen tree often reaching a height of 7 to 15 metres. Leaves posses plenty of oil glands on the lower surface. Flowers are hermaphrodite, borne at the terminals in small bunches. Each peduncle carries 3 or 4 stalked flowers at the end, the entire inflorescence being not more than 5 cm long. Each flower has a cylindrical thick ovary consisting of four fleshy sepals. Above the sepals, there are four whitish structures, petals dome shaped in appearance. After fertilization, stamens and styles fall invariably. The lower part of the flower along with the calyx develops into a fleshy, dark one seeded drupe, the sepals are reduced to triangular projections and this is popularly known as the mother clove.

Climate and Soil

Clove is strictly a tropical plant and requires a warm humid climate having a temperature of 20 to 30°C. Humid atmospheric condition and a well distributed annual rainfall of 150 to 250 cm are essential. It thrives well in all situations ranging from sea level up to an altitude of 1500 metres and also in places proximal to and away from sea. Deep black loam soil with high humus content found in the forest region is best suited for clove cultivation. It grows satisfactorily on laterite soils, clay loams and rich black soils having good drainage. Sandy soil is not suitable.

Propagation

Clove is propagated through seed which is called mother clove. The seeds become available from June to October. Fruits are allowed to ripe on the tree itself and drop down naturally. Such fruits are collected from the ground and sown directly in nursery or soaked in water overnight and the pericarp removed before sowing. They lose their viability within one week after harvest under normal conditions and hence they must be sown immediately after collection from a tree. The second method gives quicker and higher germination. Big sized seeds generally give higher per cent of germination.

Nursery practices: Beds for sowing seeds are of 15 to 20 cm height, one metre width and a convenient length. The beds should be made of loose soils and over which a layer of sand may be spread (about 5-8 cm thick). Seeds are sown at 2 to 3 cm spacing and a depth of about 2 cm. The seed beds have to be protected from direct sunlight. The germination commences in about 10 to 15 days and may last for about 40 days. In higher elevations, germination is delayed considerably, often requires 60 days. The germinated seedlings are transplanted in polythene bags (30 cm × 15 cm) containing a mixture of good soil, sand and well decomposed cowdung (in the ratio of about 3:3:1). The seedlings are ready for transplanting in the field when they are 18 to 24 months old. The nurseries are usually shaded and irrigated daily to ensure uniform stand. Vegetative propagation through grafting is not successful even on its own rootstock or on related species also.

Land Preparation and Planting

Eastern and north eastern hill slopes, well drained valleys and river banks are ideal for clove. The area selected for raising clove plantations is cleared of scrub growth before monsoon and pits of 60 to 75 cm^3 are dug at a spacing of 6 to 7 metres. The pits are partially filled with compost, green leaf or cattle manure and covered with top soil. The seedlings are transplanted in the main field during the beginning of rainy season, in June-July, and in low lying areas, towards the end of the monsoon, in September-October. Cloves prefer partial shade. It can conveniently be grown mixed with commercial crops like arecanut, coconut etc. The shade cast by these plants will provide enough protection to clove from the hot sun. In order to give a cool humid micro-climate intercropping with banana is found to be very good. Immediately after planting, mulching with available trashes is recommended.

Manuring: Clove trees are to be manured regularly and judiciously for their proper growth and flowering as given below:

Age of the plant	Cattle manure or Compost (kg)	Urea	Super Phosphate	Muriate of Potash
		g/tree		
First year	15	–	–	–
Second year	20	80	220	160
Annual increase per year	5	40	110	80
Tree of 15 years and above	50	600	1560	1250

The entire quantity of organic manures and half the quantity of fertilizers may be applied during May-June and the remaining quantity of fertilizers is applied in October-November in shallow trenches dug around the plant normally about 1 to 1.5 m away from the tree base. Apply 50 g in each of *Azospirillum* and *Phosphobacterium* one month after manuring.

After Cultivation

No intercultivation, except removal of weeds is usually done for clove. As the branches of full grown trees have a tendency to overcrowd, thinning them occasionally may keep the growth within manageable proportion. Dead and diseased shoots should be removed periodically.

Irrigation is necessary in the initial stages. In places where pronounced drought is normally experienced, pot watering is recommended to save the plants in the initial two or three years. Subsoil irrigation using 20 cm length mud tubes or bamboo tubes will be helpful to save the plants during acute summer. Although the trees can survive without irrigation, it is advantageous to irrigate the grown up trees for proper growth and yield.

Harvesting

Clove tree begins to yield from the seventh or eighth year after planting and full bearing stage is attained after about 15 to 20 years, the flowering season is September-October in the plains and December to February at high altitudes. Flower buds are produced on young flush (Figure 7).

It takes about 4 to 6 months for the buds to become ready for harvest. At this time, they are less than 2 cm long. The optimum stage for picking clove buds is indicated by the change in the colour from green to slightly

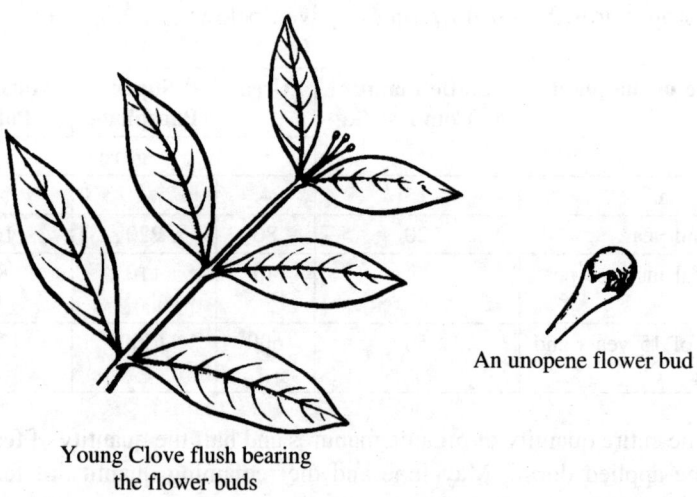

An unopene flower bud

Young Clove flush bearing
the flower buds

Figure 7 Flower Buds in clove

pinkish tinge. The matured clove buds are carefully picked with hand. Care should be taken to pick the buds at the correct time as otherwise the quality of the cured produce will be lost to a considerable extent. When the trees are tall and the clove bunches are beyond the reach, platform ladders are used for harvesting. Bending the branches or knocking down the bud clusters with sticks is not desirable as these practices affect the future bearing of the tree.

Curing: The harvested flower buds are separated from the clusters by hand and spread in the drying yard for drying. It takes normally 4 to 5 days for drying. The correct stage of drying is reached when the stem of the bud is dark brown and the rest of the bud lighter brown in colour. Well dried cloves will be only about one-third of the weight of the original. About 11,000 to 15,000 dried cloves make one kilogram.

Yield

Clove often experiences irregular or alternate bearing tendency. A well maintained full grown tree under favourable conditions may give 4 to 8 kg of dried buds. The average annual yield after 15th year may be taken as 2 kg per tree. Clove oil, the spice determining factor, is about 16 to 21 per cent in the buds. The oil contains 70 to 90 percent of free eugenol and 5 to 12 per cent of eugenyl acetate.

Plant Protection

Diseases/Pest	Symptoms	Control Measure
Leaf rot (*Cylindrocladium quinquiseptatam*)	Dark patches are seen in leaves of mature trees seedlings, which often result in rotting of whole leaves or tips alone causing severe defoliation.	Spray 0.2 percent carbendazim.
Stem borer	The pest bores into the main stem, causing death of the plant.	Spray indoxacarb 14.4 % (1.5 ml) around the hole and infect the same in to the borehole after removing the brass. Swabbing the basal region of the main stem with carbaryl 20 ml/L) and keeping the basins free from weeds.

Scale insects	Infestation is seen on leaves and tender shoots, more often in the nursery.	Spray 0.05 percent dimethoate.

4.2 NUTMEG (*Myristica fragrans* HOUTT)

Family: **Myristicaceae**

Nutmeg is native of Moluccas Island (Indonesia) and was introduced to India towards the end of the 18th century and is grown now in certain pockets of Kerala, Tamil Nadu, and Karnataka. Recent estimate showed that the total area under nutmeg in India is 17,730 ha, producing 12,620 tonnes annually. Kerala has nearly 98 % of the area. Over 50 percent of the world exports originate from Indonesia and Greneda is the second largest exporter of nutmeg and mace. India is both exporting and importing the nutmeg and mace .Annually about 720 tonnes of nutmeg (valued Rs. 3575 lakhs) and 560 tonnes of mace (valed Rs. 4937 lakhs) are imported while we are exporting about 3645 tonnes of nutmeg and mace (Valued Rs.26095 lakhs). India should therefore step up its production to meet its own requirement.

They are important to confectionary, culinary and pharmaceutical industries. Nutmeg and mace also yield 7 to 16 and 4 to 15 percent of oil respectively. This oil is used for flavouring food products and liquors and also in perfumery industries. Oleoresin of nutmeg and mace is used almost entirely in the flavouring of processed foods. The nutmeg butter prepared by the expression of fat from the ground, cooked nutmeg is mostly used to impart spicy odour to the perfumes. The pericarp is used for making jams, jellies and pickles.

Botany

Nutmeg is a densely foliated evergreen tree growing to a height of 20 metres and above. The trees are normally dioecious, producing male and female trees separately. However, it is not uncommon to meet with bisexual trees. The male flowers are solitary and borne in the leaf axils. Pollination is effected by wind and insects. The fruit is a fleshy drupe, spherical or slightly ovoid in shape and pale yellow in colour with a longitudinal groove in the centre. When fully ripe, the fleshy rind splits open along the groove into two halves, exposing the bright red aril or 'mace' of commerce and brownish black seed inside. The other related species are *M. fatua* var.

magnifica, M. malabarica, M. beddomi and *Knema andamanica* however they donot produce mace of economic importance.

Climate and Soil

Nutmeg thrives well in warm humid conditions in locations with an annual rainfall of 150 cm and more. It grows well from sea level up to about 1300 metres above MSL. Areas with soils of clay loam, sandy loam and red lateritic are ideal for its growth. Both dry climate and water logged conditions are not good for nutmeg. The soil and climate prevailing in the hill slopes of Western Ghats and Eastern Ghats (from 700 to 1300 m) are ideally suited to grow nutmeg.

Varieties

S. No	Variety	Pedigree / parentage and plant type	Yield per tree	Salient features
1.	Konkan Sugandha	Single plant selection from local seedling population	200-526 fruits/ tree	Adaptable in Konkan region. Tree canopy is conical and compact.
2.	Konkan Swad	Selection from nutmeg seedlings from Ratnagiri dist.	761.38 fruits/ tree	Adapted to Konkan region with warm, humid conditions as well as shade provision. Canopy erect, conical shape Contain 39.8 % essential oil in seed and 10.9% in mace. No incidence of pest & disease are noticed.
3.	Konkan Shrimanti	A single plant selection	9.3kg dried nut/tree/year, 1.9 kg of dried mace/ tree/year.	Bold nuts, thick mace
4.	Viswahree	Clonal selection from elite germplasm	1000 fruits (1.33 kg mace, 9 kg dry 3 nuts,	Bushy and compact canopy, low incidence of fruit rots. Nut recovery 70.0%, mace recovery 35.0%

				oil 7.14%, mace oil 7.13%, oleoresin 2.48% and mace 13.8% respectively, butter 30.9%, myristicin, 12.48% and mace 20.03% respectively
5.	Kerala Shree	Clonal Selection from a Farmer's Field	2000 fruits / graft at 10th year	Fresh fruit weight (75-100 g), fresh weight of seed (13-16 g), dry recovery of nut (70%) and aril, fresh weight of aril is 4.5-6.0 g.

N.B **1.** Released from Regional Fruit Research Station, (BSKKU), Vengurla, Maharastra,**2.** Released from Regional Coconut Research Station, Bhatye (Dr. BSKKU), Ratnagiri, **3.** released from Dr. Balasaheb Sawant Konkan Krishi Vidyapeeth, Dapoli, **4 and 5-** released from Indian Institute of Spices Research, Calicut, Kerala.

Propagation and Nursery Techniques

Nutmeg trees are usually propagated through seed. Large sized tree-burst fruits from high yielding trees are harvested or collected as and when they fall down. The fleshy rind of the fruit as well as the mace covering the seed is removed before sowing. The seeds should be sown in specially prepared nursery beds immediately after collection. If there is delay in sowing, the seeds may be kept in trays filled with moist sand or stored in poly bags or moss which keep them viable for 15 days.

Regular watering is necessary for good germination. The germination may commence from about the 30th day and lasts up to 90 days after sowing. About 20 days old sprouts are transplanted to polythene bags containing a mixture of good soil, sand and well decomposed cow dung (3:3:1). About 18 to 24 months old seedlings are used for transplanting in the field. Air layering, inarching and budding have also been attempted with varying degree of success, but the epicotyl grafting technique developed by NRCS, Calicut is found to be promising now (Fig. 8).

In this method, the selected root-stock of first leaf stage should have a thick stem (diameter of 0.4 cm or more) with sufficient length to give a cut of 3.0 cm long. The scions with 2 to 3 leaves, collected from the high yielding tree can be used for grafting purpose. The stock and scion should have approximately the same diameter. A 'V' shaped cut is to be made in the stock and a tapered scion (wedge shape) is fitted carefully into the cut.

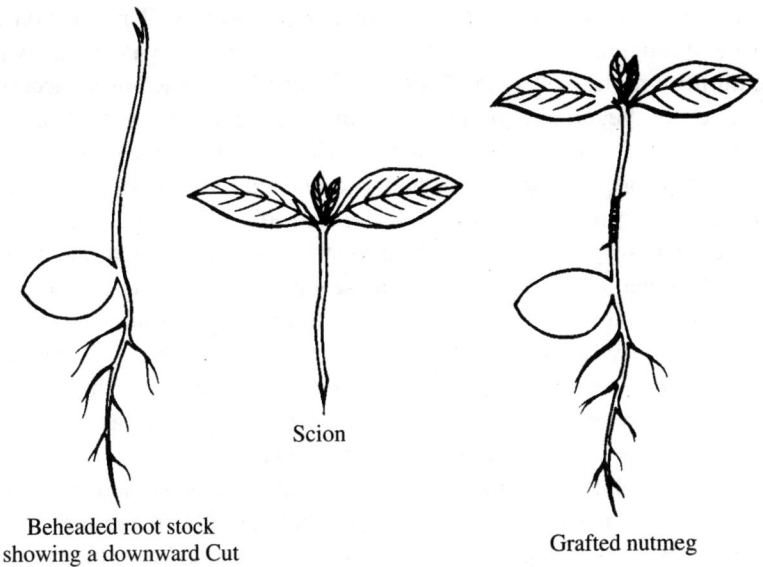

Scion

Beheaded root stock
showing a downward Cut

Grafted nutmeg

Figure 8   Steps in epicotyl grafting in nutmeg.

Tying may be done with 300 gauge polythene bags, containing coir dust as medium. The polythene bags are used to tie the grafted portion to prevent drying of scion and should be kept in a cool shaded place protected from direct sunlight. After one month, the bags can be opened and those grafts showing sprouting of scions may be transplanted in bags containing a mixture of soil, sand and cow dung in the ratio of 1:1:1 and kept in shade. The polythene bandage covering the grafted portion can be removed after three months of transplanting. During grafting, precautions should be taken to prevent wilting of scions and to complete the grafting as soon as possible after detachment of shoots. Since the exudates from cuts pose a problem of covering the cut surfaces in the grafting process hindering union, the rootstocks are to be prepared earlier than the scions, so that by the time the scion is ready exudation might have ceased in the rootstock.

Straight growing shoots (orthotropic or chupon) should be always used as scion to get normal shaped nutmeg trees. If the plagiotropic shoots (lateral branches) are used, a shrubby spreading plant will result which need special efforts to get converted into a normal growing tree, the effort may not always be successful.

Recently, **'green chip budding'** is recommended to produce orthotropic plants through budding. This method of budding is superior to other methods of budding because here the bud is removed with the cambial tissue and the union is faster resulting in higher success percentage. This method can be done throughout the year but the success percentage is higher when done during July to November. This method can be adopted as a successful technique for vegetative propagation of nutmeg using orthotropic and plagiotropic buds from high yielding female trees through this technology is best suited for production of orthotropic budded plants due to the limited availability of orthotropic buds. The orthotropic shoots existing on the high yielding trees are pruned for production of new orthotropic shoots. The newly emerged shoots with six to seven buds are used as bud wood as green buds are used for budding in this technique. Bud wood has to be collected early before it becomes too hot, on the same day of budding. In order to prevent moisture loss from the bud wood/stick, the collected bud wood is wrapped in a moist paper placed in a plastic bag as soon as they are harvested. Make sure that scion wood is also clean of soil and free of disease and pest. Unlike in other budding methods, here the bud is removed with a part of the wood and also half portion of the leaf lamina is retained. By making appropriate horizontal and vertical cuts the bud chip can be removed.

Planting: Seedlings are transplanted in the main field when they are 12 to 18 months old. Recommended spacing is 8 × 8 metres. Pits of 60cm cube are dug and filled with compost and top soil. Seedlings are planted carefully in the centre of the pits during rainy seasons. Young plants are provided with artificial shade and irrigated during summer months. The basins are mulched with locally available trashes. As nutmeg requires light shade especially at early stage, fast growing shade plants like *Erythrina indica, E.lithosperma or Gliricidia maculata* or bananas are planted in between them a few months prior to planting. As nutmeg plants grow, these shade plants may be thinned out. Though not essential, irrigation is advantageous to grown up trees for getting better yield.

Nutmeg can be advantageously grown mixed with coconut and areca-nut. When coconut is planted at a spacing of not less than 8 × 8 m one nutmeg can be planted at the centre of four coconut plants. In arecanut plantations raised with a spacing of 2.7 m × 2.7 m, nutmeg can be planted at every third row of arecanut so that within the square formed by four nutmeg plants there will be nine arecanut seedlings. In coffee based mixed system also, nutmeg can be planted at a spacing of 7-8 m.

Manuring and Intercultivation

Nutmeg requires heavy manuring for proper growth and yield. Farm yard manure or compost at the rate of 10 kg per plant may be applied during the first year of planting. The quantity should be increased every year so that a well grown tree of about fifteen years and above may receive 50 kg organic manure. During the first year, fertilizers to supply 20 g nitrogen (N), 18 g phosphorus (P_2O_5) and 50 g potash (K_2O) are applied per plant. The fertilizer doses may be increased gradually every year till a well grown tree of fifteen years or more receives about 500 g nitrogen (N), 250 g phosphorus (P_2O_5) and 1000 g potash (K_2O) annually.

The fertilizers are applied in two equal split doses, first in May-June along with organic manure and the other in September- October. Shallow trench is dug around each tree at a radius of 1 to 1.5 m away from the trunk for applying manures and fertilizers and covered after application of manures.

As nutmeg has a very shallow root system, trenches should not be deep and no intercultivation is usually done very close to the trunk. However weeding is done periodically.

Harvesting and Post Harvest Processing

Seedling trees start bearing in 7 to 8 years while grafts start bearing in 4 to 5 years. They attain full bearing stage after 15 to 20 years and may yield up to 60 years. Nutmeg trees flower throughout the year with a peak in certain months. Hence, though fruits are seen throughout the year, the peak season of harvest is from June to August. The fruits take 9 months from flowering to harvest. When the fleshy rind of the nut splits open, the fruits are fully ripe for picking. They are either plucked from the tree or are allowed to drop on to the ground and then collected. After removal of the outer fleshy rind, mace is detached from the seed shell by hand and flattened out. It is then allowed to dry slowly in the sun for ten to fifteen days. During drying the mace gradually becomes brittle and horny and attains a yellowish brown colour.

The seeds are dried separately for 4 to 8 weeks either in the sun or in artificial heat until the kernel rattles inside the shell which is then broken open with a wooden mallet and the kernel is taken out. The fleshy pericarp can be used for making pickles, jams and jellies.

IISR, Calicut has standardized drying of nut and mace using hot air oven. Freshly harvested mace can be blanched in water at 75°C for two minutes to retain the scarlet / red colour. This is followed by hot air drying at 55° – 65°C which takes three to four hours for drying to a moisture level of 8–10 per cent. However nut can be dried in 14–16 hours using hot air technique. Reduction in mould and fungal population in dried mace due to blanching is also reported in this method besides retention of colour. Solar drying technique is also employed in which the energy from the sun is used to heat a stream of air which inturn flows by natural or forced connection, through a bed of commodity to be dried. This method ensures quality enhancement and reduction in drying time.

Whole nutmegs are grouped under three broad quality classification: sound, substandard and distilling. Sound export nutmegs are generally graded in Indonesia by size into approximately 170/180 or 240/250 nuts/kg with large uniform nutmeg or 'ABCD' which is an assortment of various sizes. Substandard nutmegs are traded as 'sound, shriveled', which in general have a higher volatile content than mature sound nutmeg and are used for grinding, oleoresin extraction and oil distillation, and BWP (broken, wormy and punky) which are used mainly for grinding, as volatile generally does not exceed eight percent. Distilling grades of nutmeg are of poor quality: 'BIA' or 'ETEZ' with a volatile oil content of 8 per cent to 10 per cent and "BSL" or "AZWI" which has less shell material and a volatile oil content of 12-13 per cent. Mace is classified as whole pale mace, No.1 broken mace, selected, unasserted or siftings (Indonesia) and as whole, broken blades or siftings (Grenada). Nutmegs are usually packed in double layered linen, jute, sisal or polythene bags. If other packing materials are used, care must be taken to avoid materials which might lead to 'sweating' and mould development. Adulteration is common in the nutmeg trade. *M. fragrans* is adulterated with *M. argentea*, *M. malabarica* and which can be identified by their poor quality. The mace from *M. argentea* is imported as Papuan nutmeg from Papua New Guinea; *M. malabarica* is traded as Bombay nutmeg from India and *M. otaba* as Otaba nutmeg. Trade of wild nutmeg exists and they are marketed as long, female, Papua, Guinea, or Norse nutmeg.

Yield

Individual nutmeg fruit weighs on an average 60 g of which the seed weighs 6 to 7 g, mace 3 to 4 g and the rest pericarp. A fully grown tree under normal conditions may produce on an average 2,000 to 3,000 fruits per year. On the assumption that about 65 percent of the trees in a garden are female and are optimum yielders, about 800 kg nutmeg and 100 kg mace can be expected per hectare. With 360 grafts per ha, nut yield of 7000 kg can be harvested after 10th year.

Sex in Nutmeg

The main problem facing the nutmeg cultivation is the segregation of seedlings into 1:1 ratio for male and female trees resulting in 50 percent of unproductive trees. The sex identification at seedlings stage itself is very difficult. Hence to overcome this problem, planting of grafts using scion from female tree is advocated now. For every 10 female trees, one male tree has to be retained. In the established plantations, the unproductive male plants can be converted into productive female trees by top working. After the first flowering, excess of male plants (7 to 10 years old) are cut back at a convenient height leaving few branches below. Just below the cut end, patch budding or side grafting or cleft grafting (Figure 9) is done during the end of rainy season. Orthotropic shoots (scion) from female trees are alone used.

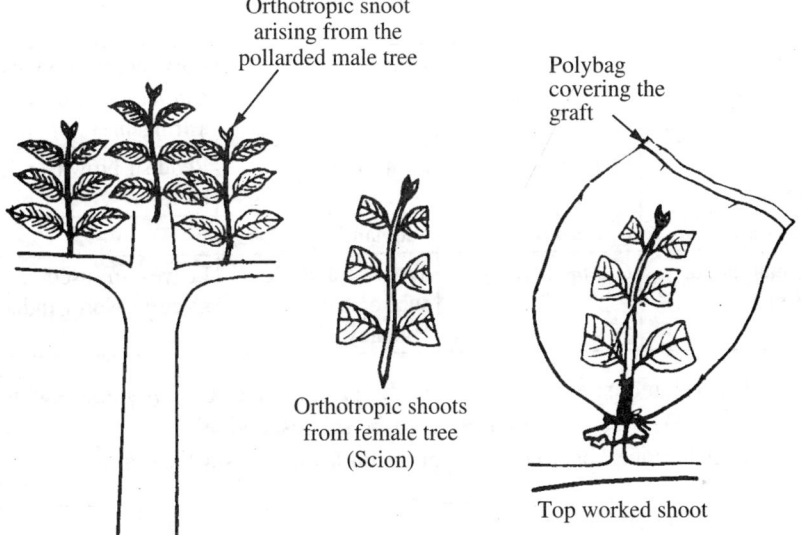

Figure 9   Steps in top working of male nutmeg tree

Plant Protection

Diseases	Symptoms	Control measure
Die back (*Diplodia natalensis*)	Drying up of mature and immature branches from tip downwards.	Remove the infected branches; smear Bordeaux paste in the cut end of the branches.
Thread blight (*Marasmius sp.*)	Blighting of the affected stems or leaves	Spray 1.0 percent Bordeaux mixture.
Fruit rot (*Diplodia natalensis* and *Phytophthora sp.*)	Splitting of immature fruits and subsequent rotting of mace.	Spray 1.0 percent Bordeaux mixture when the fruits are half-mature.

4.3. CINNAMON (*Cinnamomum verum* P. resl.)

Family: **Lauraceae**

True cinnamon also called as 'Ceylon cinnamon' or 'sweet wood', one the oldest known spices, is the dried bark of *Cinnamomum verum* (Syn. *C. zeylanicum)*. Cassia, popularly known as, 'Chinese cinnamon', 'false cinnamon' or 'bastard cinnamon' also one of the oldest known spices, is obtained from the bark of *C.cassia*. The other important economic species of *Cinnamomum* are given below:

Common name	Botanical name	Commonly grown in	Remarks
Chinese cassia	C. cassia (syn. C.aromaticum)	China	Bark sweet, aromatic and leaf slightly astringent.
Indonesian cassia	C. burmannii	Java, Sumatra	Reddish brown barks.
Saigon cassia	C. loureirii	Vietnam	–
Indian cassia (Tejpat)	C. tamala	Tropical and subtropical Himalayas	Leaves are used as a spice in North India.

Other related species are *C. malabaricum* and *C. camphora* which contain 0.11-0.33 % oil and 1.50-2.90 % oil respectively.

The differences between cinnamon bark and cassia bark are:

Cinnamon bark	Cassia bark
Bark is smooth and thin	Bark is coarse and thick
Bark powder is tan in colour	It is reddish brown in colour
It has relatively less intense aroma	It has more intense aroma
Essential oil content is less (0.5 to 2.0%)	Essential oil content is high (1 to 4.5%)
Eugenol is the main constituent in the leaf oil	Cinnmaldehyde is the main constituent in the leaf oil.
Flavour is good	Not so delicately flavoured.

Sri Lanka followed by Sychelles are the largest producer of cinnamon bark with best quality. In India, it is grown in Kerala, Tamil Nadu and Karnataka. Indonesia, China and Vietnam are the largest producers of cassia and it is not grown in India but collected from forest areas of North eastern states mainly from Assam, Meghalaya and Manipur. The area under cultivation in India including Tejpat is about 2950 ha producing about 5000 tonnes of cassia and cinnamon but our production is more warranting import of nearly 12000 tonnes of cassia valued around Rs.8000 lakhs.

Climate and Soil

Cinnamon is a hardy plant. It tolerates a wide range of climatic and soil conditions. It comes up well from sea level up to an elevation of about 1,000 m. Since it is mostly raised as an unirrigated crop, an annual rainfall of 200-250 cm is considered ideal for the crop. In the West Coast of India, the tree is grown on laterite and sandy patches with poor nutrient status.

Varieties

Many centres have developed improved varieties as indicated below:

S.No	Variety	Pedigree / parentage and plant type	Yield t/ha (fresh)	Salient features
1.	YCD.1	Clonal selection from OP seedlings progenies of Sri Lankan type.	Dry bark yield of 360 kg /ha (143.9g/dry bark/plant)	Good bark recovery, adapted to wide range of soil and rainfed conditions. Bark oil 2.8%, leaf oil 3.0%, bark recovery 35.3%.
2.	PPI (C)-12	Selection from OP seedlings progeny introduced from Sri Lanka (Seed supplied from IISR)	Bark yield of 980 kg/ha.	Bark, higher oil recovery from the bark (2.9%) and leaf oil recovery of 3.3%, bark recovery 34.22%. Suitable for cultivation in high rain fall zones and hill regions of Tamil Nadu at an altitude range of 100-500m MSL.
3.	IISR Nithyasree	Clonal selection from OP seedling progeny	200 kg dry quills	Good regeneration capacity, Bark & leaf oleoresin contents are high, good bark recovery with good aroma and taste. Bark oil 2.7%, leaf oil 3%, Bark Oleoresin 10%,

4.	IISR Navashree	Selection from Op seedling progeny of Sri Lankan collections	200 kg	High quality line, good bark recovery with good aroma and taste, grow well in plains and higher elevations. Bark oil 2.7%.Leaf oil 2.8%, bark oleoresin 8.0%, and Bark recovery 40.6%.
5.	Konkan Tej	Seedling selection from progenies of Sri Lankan accessions	Bark yield of 334 g/plant	Superior qualities with 3.2% bark oil with bark recovery 29.16%, leaf oil 2.28%, bark recovery 51.78%.
6.	Sugandhini (ODC -130)	Single tree selection from Wynad local collection. A Sri Lankan type	640 Kg	Recommended for cultivation for leaf oil production, bark oil 0.94%, leaf oil 1.6% bark recovery 51.0%
7.	RRL (B) C-6	Selection from germplasm collection – OP seedling progenies.	250 Kg bark/ ha	High quality, sweet and pungent bark, leaf oil 0.8% spreading branching nature.

N.B 1. Released from Hort. Research Station (TNAU), Yercaud, **2.** Released from HRS (TNAU), Pechiparai, **3 and 4.** Released from Indian Institute of Spices Research, Calicut **5.** Released from Regional Coconut Research Station, Dr. BSKKV, Vengurle, Maharashtra, **6.** Released from Aromatic and Medicinal Plants Research Station (KAU), Odakalli, Kerala **7.** Released from Regional Research Laboratory, Bhubaneshwar, Orissa.

Nursery Practices

The most common method of propagation of cinnamon is through seed, though it can be propagated by cuttings as well as layers. Cinnamon generally flowers in January and fruits ripen during June-August. The fully ripened fruits are either picked up from the tree or the fallen ones are collected from the ground. Seeds are removed from the fruits, washed free of pulp and sown immediately. The seeds are sown in beds or polythene bags containing a mixture of sand, well rotten cattle manure and soil (2:1:1). Germination occurs within 15 to 20 days. Six months old seedlings are transferred to mud pots or polybags.

Planting

Seedlings are transplanted in the main field when they are about 12 months old at a spacing of 3 metres on either way. The planting is done in pits of 60 cm cube which are dug earlier and filled with compost and top soil. The most appropriate time for planting is June-July to take advantage of the South West monsoon. In the early period, artificial shade is provided. Seedlings grow well in the open as well as under partial shade.

Manuring and intercultivation: Systematic manuring is rarely practised in India for this crop. The common practice is to mulch the plant bases with the trash obtained during weeding. One earthing up is given at the time of shoot collection in June. For better growth, it is essential that the plants are manured systematically. Fertilizers to supply 20 g nitrogen (N), 18 g phosphorus (P_2O_5) and 25 g potash (K_2O) are given per seedling in the first year and this dose is doubled in the second year. Every year the dose of fertilizers is increased gradually so that grown up plant of 10 years and above may receive 200 g nitrogen (N), 180 g phosphorus (P_2O_5) and 200 g potash (K_2O). Cattle manure or compost at the rate of 20 kg per plant may also be applied. The entire organic manure and half the dose of fertilizers are applied in shallow trenches dug around the plant once in May-June and the remaining fertilizers are applied in September-October.

Harvesting and Processing of Cinnamon

Cinnamon tree may attain a height of about 10 to 15 metres, but in cultivation, it is generally coppiced or cut back periodically. When the plants are two years old, they are coppiced during June-July to a height of about 12 cm from ground level. The stump is then earthened up. This operation encourages the development of side shoots from the stump. The coppicing is repeated for every side shoot, developing from the main stem during the succeeding season, so that the plant will assume the shape of a low bush of about 2 metres height and a bunch of canes suitable for peeling would grow up in a period of about four years. Regular peeling operations could be commenced from the fourth or the fifth year, depending upon the extent of development of peeler shoots.

There are two regular cutting seasons in South India, which more or less synchronise with the two monsoons. The appropriate time for cutting the shoots for peeling is determined with reference to the circulation of sap between the wood and the corky layer. A test cut can be made on the stem with a sharp knife to judge the suitability of the time of peeling. If the bark separates readily, the cutting can be commenced immediately. The stems are cut close to the ground when they are about 2 years old, as straight as possible, 1.0 to 1.25 m length and 1.25 cm thickness. Such shoots are bundled after removing the leaves and the terminal shoots.

Cutting is followed by scraping and peeling operations. It is done by using a specially made knife, which has a small and round end with a projection on one side to facilitate ripping of the bark. A new knife fabricated at TNAU Horticultural Research Station, Thadiankudisai (Fig. 10) facilitates easy extraction of bark and it saves 30 to 40 per cent time in extraction. The rough outer bark is first scrapped off. Then with a brass rod, the scraped portion is polished to facilitate easy peeling. A longitudinal slit is made from one end to the other. Then working the knife between the bark and the wood, the bark is ripped quickly. The shoots cut in the morning are peeled on the same day. The peels are gathered and kept overnight under shade. They are dried first in shade for a day and then in the sunlight for four days. During drying, the bark contracts and assumes the shape of quill. The smaller quills are inserted into larger ones to form compound quills.

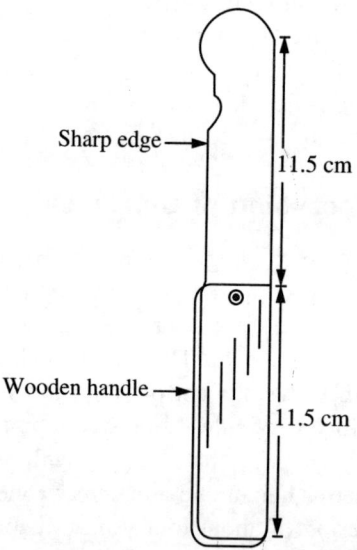

Sharp edge ——

11.5 cm

Wooden handle ——

11.5 cm

Figure 10 Cinnamon bark peeling knife

Grades: The quills are graded from 00000, being the finest quality, to O the coarsest quality. The small pieces of the bark, left after preparing the quills are graded as quillings. The very thin inner pieces of bark are dried as **featherings.** From the coarser canes, the bark is scraped off instead of peeling and this grade is known as scraped chips. The bark is also scraped off without removing the outer bark and is known as **unscraped chips.** The different grades of bark are powdered to get **Cinnamon powder.**

Yield

Yield from one hectare of cinnamon plantation, which is in full yielding stage, is about 200 to 300 kg of dried bark. Leaf oil and bark oil of cinnamon can be also distilled from the dried leaves and bark respectively. Barks having purple coloured young flushes yield more oil than green flushes type. The oil has a pungent odour and hot taste and contains 70 to 80 per cent of eugenol with traces of cinnamic aldehyde. The leaf oil and bark oil are used commercially in many pharmaceutical, perfumery and other industries.

Plant Protection

Diseases/Pests	Symptoms	Control measure
Diseases Pink disease (*Corticlum javanicum*)	A pale pinkish white crust on the stem, destroying the cork layer and causing death of the twig or shoot.	Spray 1 % Bordeaux mixture.
Seedling blight (*Diplodia sp.*)	Blighting of the stem of young seedlings.	Drench 1 % copper oxychloride.
Leaf spot (*Gloeosporium sp.*)	Dark brown, sunken spots	Spray any copper fungicides (0.1 %)
Pest Cinnamon butter fly (*Chilasa clytia*)	Caterpillars feed voraciously on young leaves and cause defoliation.	Spray flubendiamide 39.35 % SC (0.6 ml/L)

4.4. ALLSPICE (*Pimenta dioica* (L.) Morr.)

Family: **Myrtaceae**

The allspice of commerce is the dried immature fruits of the tree *Pimenta dioica*. It is indigenous to West Indies. Jamaica is the main producer of allspice. Its flavour is said to resemble a blend of cinnamon, clove and nutmeg. There are a few trees available in the Mahendragiri hills of Nagercoil area, Ambalavayal in Kerala and Kallar and Burliar of the Nilgiris.

It is a small evergreen tree; flowers are white and branch trichotomously in the axils of upper leaves. Flowers are structurally hermaphrodite but functionally dioecious. Stamens numerous, above 100 in barren trees and

50 in bearing ones. It flowers during March-June and matures in 3 to 4 months after flowering. Fruit is a two seeded berry. Male trees flower earlier.

Cultural Aspects

It is raised through seeds. The seeds can be stored as ripe berries after collection without extracting seeds upto three weeks. But viability gets reduced slowly after a period of three weeks and is lost completely after nine weeks. The seeds are sown in raised beds of 15 to 20 cm high, one metre width and convenient length. The beds may be made of loose soil-sand mixture having sand in the top layer. The seeds germinate in 15 days and continue up to 40 to 45 days. The seedlings can be transferred to bags 3 weeks after their emergence above ground level. Six months old seedlings are planted at a spacing of 6 m either way. There is no manurial schedule available for allspice and hence the schedule recommended for clove can be followed for this. Plants start flowering when they are 7 to 10 years old, but the peak harvest is obtained from 15 to 20 years.

Harvesting and Processing

The berries are harvested when fully grown, but still green. They are sun dried and are frequently turned down with ladles. The process of curing lasts 3 to 12 days, till the berries are completely dried. Curing is complete when the berries become crisp and produce a metallic sound if shaken. A tree yields 50 to 60 kg of dry berries.

The Pimento berry oil is distilled commercially in Jamaica. The oil yield from berries varies from 3.3 to 4.5 per cent and its eugenol content is 65 to 66 per cent.

The pimento leaf oil is also distilled only in Jamaica from the leaves removed during the harvesting of the berries. Oil yield from dried leaves ranges from 0.7 to 2.9 per cent while fresh leaves yield 0.35 to 1.25 per cent. It is a flavouring agent in culinary preparations like soups, sauces, ketchups, pickles, meat, and curry powder and in bakery items. In liquor industry also, it is used as a favourite ingredient.

4.5. TAMARIND (*Tamarindus indica* L.)

Family: Caesalpinaceae

Tamarind is native to tropical Africa and is now widely planted and naturalised everywhere in the tropics. It grows widely in the tropical and subtropical regions of subcontinental and is also planted along avenues, in parks etc. The area under tamarind cultivation in India is 58.11 ha and production is 201.82 tonnes annually. The fruits are used for various culinary purposes all over India as they contain tartaric acid (8%). The seeds give a gum used for binding and the wood has a good timber value.

It is a medium sized tree with a short strong trunk. The leaves are pinnately compound with 10-20 pairs of small leaflets. The flowers are small, scented, and attractive with yellow and red coloured. Pods are flattened and vary considerably in size and shape. When ripe, the fruits are stiff and brittle. The most valuable part of the tamarind fruit is the brown and sweetly acid pulp. There are reddish-pulp types which are considered the best.

Tamil Nadu Agrl. University (H.C.R.I, Periakulam) has released one improved tamarind variety PKM.l-Tamarind by clonal selection. It has a high pulp recovery of 39%, pods are less fibrous and pulp is also very sweet. It yields 250 kg from ninth year. Another local selection viz Urigam, a long podded type, measuring more than 20 cm in length, are also becoming popular among the growers. The grafts of this selection also start bearing from third year onwards. The other selections identified in Karnataka are DTS 1) which has sweet red type, slightly susceptible to pod borer and Yogeswari (Red Type) and in Maharashtra are Prathisthan and Selection No263.

Climate and Soil

Tamarind can thrive in tropical and subtropical climate excepting in places experiencing frost. Similarly, it is not exacting in its soil requirement. It thrives even in sodic and saline soils, ravines and degraded land.

Propagation: It is generally propagated through seeds. Vegetative propagation through whip cum inarching method gives more than 90 percent success. Patch budding on 9 months old saplings also gives more than 90 percent success.

Planting: It may be planted from June to November in South India and a spacing of 10 ×10 m may be adopted to accommodate 100 plants, per ha. A pit size of 1 m^3 may be dug for planting and is refilled with topsoil and farm yard manure for good establishment of the saplings.

Aftercare: Regular watering should be given till the plants establish well in the field. As it is a dry land crop, it is not manured generally. But it responds to organic manure. In the case of grafts or buddlings, the rootstock sprouts should be periodically removed. Intercrops may be raised in the alley spacing up to 4 or 5 years.

Harvest: The seedlings take 8 to 10 years for fruiting and the grafts take 4 to 5 years for fruiting. It is not uncommon that trees with more than 100 years are yielding normally. There is a tendency of alternate bearing in this crop. Fruits are harvested from December to April.

Yield: Though the graft starts yielding from fourth or fifth years, good yield of 250 kg per tree is obtained from ninth year in PKM-1 variety.

Plant Protection: Among the pest, mealy bugs and scales are often causing serious concern especially in the nursery and can be controlled by spraying dimethoate 30 EC (2 ml/L) . Powdery mildew sometimes infects the leaves and can be controlled by spraying 0.1 % Karathane.

4.6. CURRY LEAF (*Murrya koenigii* Spreng.)

Family: **Rutaceae**

Curry leaf plays an important role as a condiment in the culinary preparation of South Indian dishes. A plant of homestead gardens has recently gained importance as a commerical crop and is cultivated in field scale in Coimbatore, Periyar, Madurai, Salem and Trichy districts of Tamil Nadu and in Dharward, Belgaum and Uttara Kannada of Karnataka State.

Climate and Soil

It does not exact a specific climate and can come up in dry climate too. In places where minimum temperature goes below 13°C, the growth of shoot will be slightly affected. Though it can be cultivated in most of the soil types, it comes up well in light textured red soils.

Varieties

There is no named variety available and farmer prefers local varieties which have pink midrib. University of Agricultural Science, Dharwad has recently released two cultivars viz DWD -1 and DWD - 2 which have an oil content of 5.22 and 4.09% respectively. Both the varieties are having strong aroma. DWA-1 (Suvasini) is a clonal selection from root suckers and the plant has dark green highly aromatic shining leaves. It is sensitive to low temperature in winter and hence bud burst is poor during winter. The leaves have an oil content of 5.22%.

Preparation of Field and Planting

Main field should be ploughed repeatedly. Normally, a spacing of 90 to 120 cm is followed on either side. High density planting (0.75 m on either side) is also followed. One month before planting, pits of 30 × 30 × 30 cm dimensions are dug out and are filled with top soil mixed with well decomposed farm yard manure at the time of planting. Healthy seedlings are planted in the centre of the pits. Then long furrows are formed connecting all the pits to facilitate easy irrigation.

The seedlings are irrigated once in five to seven days up to three years and once in 15 days afterwards.

This being a perennial crop, the field should be kept free of weeds. It also responds to irrigation. Under commercial scale, drip irrigation is being practiced in certain parts of Coimbatore district. Normally curry leaf plants are not fertilized with inorganic source by the farmers. However, for a better growth and yield, each plant may be fertilized with 20 kg of Farm yard manure besides 150:25:50 g of NPK per year. Poultry manure @ 12 t/ha and foliar spray of panchagavya (3 %) increases the leaf yield.

Curry leaf should not be allowed to grow beyond 1.20 m. In order to encourage more branches, when they are about 6 months old planting, they may be pruned at 15 cm height from ground level. 2-3 harvest can be made thereafter and again they are to be pruned at 15 cm height to maintain the bush height. In this way, we can have 3-4 harvests per year and the plantation can be retained upto 15-20 years. In well maintained field, 8-10 tonnes of curry leaf can be harvested per year.

Many pests such as mealy bugs, aphids, leaf rollers are known to attack the curry leaf. Control such as measures involve use of safe chemicals or neem products as the leaves are directly used in culinary preparations.

Harvesting and Yield

The plants may be trained and pruned to maintain a bush of 1.0 m height. At the end of first year, the crop comes to first harvest. The expected yield from one hectare is given below:

Age of plantation	Yield of leaves (kg/ha)	Interval of harvest
First year	400	Once at the end of first year
Second & third year	2000 to 2200	Once in four months
Fourth year	2500	Once in 3 months
Fifth year onwards	3500 to 5000	Once in $2\frac{1}{2}$ to 3 months

The plantation can be kept for 20 to 25 years depending upon the management practices followed.

Plant Protection

Aphids attack the plant when the plants are in vegetative stage and spraying of dimethoate at the rate of 2 ml/L of water is recommended. Leaves from such sprayed plants should not be harvested for 10 days lest residual toxicity of the pesticide.

Leaf spots sometimes occur and spraying carbendazim at the rate of one gram per litre is recommended.

4.7. GARCINIA (*Garcinia cambogia* L.)

Family: **Guttiferae**

Garcinia cambogia, Syn. G.gummigutta popularly called Kudanpuli in Malayalam produces an acidic fruit which is extensively used in Kerala in culinary preparations involving fish. The processed rind is an excellent substitute for tamarind in cooking.

This evergreen dioecious tall growing tree, though native of Western Ghats of Kerala from 1200 to 2000m above MSL is now found in all parts of Kerala in backyards of homesteads. It flowers from February to April and fruit ripens during June- July. The fruits are ovoid or spherical in shape and may vary in size weighing 50 to 180 g and the fruit rind is grooved into 7 to 10 segments.

In Maharashtra, one improved clone viz Konkan Amrutha has been identified. Its yield potential is 138.28 Kg fruits / tree and the fruits attractive, early maturing type, apple like shape, 81.72% moisture, 9.08% TSS, 2.41% reducing sugar, 4.52% total sugar, 5.12% acidity with longer shelf life.

Fruits abscise on ripening and are collected and the seeds and rinds are separated. The rinds are sun dried for one or two weeks till they attain a coal black colour and characteristic acrid taste. These rinds are preserved by rubbing with salt and cocount oil. They contain 10.6 percent of tartaric acid.

Another related species, *Garcinia indica,* popularly known as punam-puli in Malayalam or Kokan in Hindi or Murgal in Tamil is a solitary tall tree grown in the coast of tropical south western part of India. Its fruits are round, purplish plum like containing light seeds. Its skin is used in cooking as a souring agent, as it has a sour and rather salty taste. Kokum syrup, Kokum Agal and Kokum Amsul are normally prepared. Many value added products like kokum rind powder, kokum honey, kokum wine are also developed now. Hydroxy Citric Acid (HCA) and anthocyanins are also extracted from this fruit now which have lot of nutraceutical properties.

4.8. BAY LEAF (*Laurus nobilis* L.)

Family: **Umbelliferae**

It is a temperate evergreen tree native of Mediterranean region, but grown as a bush in large containers. It is propagated through cutting or layering. Its leaves are harvested by manual picking, dried in shade, used extentsively in non-vegetarian dishes in Western countries and are mainly used in flavouring vinegar, pickle, soups, meat fish and sauces. The leaves have a pleasant odour with a bitter taste and contain 3 percent oil.

5

Herbal Spices

Herbal spices are defined as those herbs whose leaves are mostly used for seasoning the dishes. These herbal spices have medicinal and cosmetic uses also. It is estimated that about 39,000 tonnes of herbal spices are produced throughout the world and among them 31,000 tonnes are sundried and the balance is dehydrated or freeze dried. They are largely used in America and Europe. Our country annually requires about 200 tonnes of these spices, but 65 tonnes are alone produced in India. The bulk consumption of these group of spices is accounted for by the four metropolitan cities viz., Delhi, Bombay, Calcutta and Madras and a few in other cities like Raipur, Bangalore, Visakhapatnam and Srinagar where foreign tourists visit in large number.

These spices can be grown throughout the year in some of the hill stations of South India, North East India and valleys of Himalayas and only during winter months in the plains of North India. Important herbal spices which hold promise for cultivation are given below:

Common name	Botanical name	Important cultivation hints
1. Basil, French basil or sweet basil	*Ocimum basilicum* (L) Fa: Labiatae	Tropical perennial herb, seed propagated, harvesting is through stripping the leaves, culinary uses, yield 0.5% oil on steam distillation and is used for flavouring and in perfumery, fresh leaves are also used in pharmaceutical preparations.
2. Horseradish	*Armoracia rusticana* Fa: Cruciferae	Hardy temperate perennial, propagation through root cuttings, plants are lifted when 3 years as old, root highly pungent used mostly as a condiment and also in preparation of beef, fish etc., Tender leaves are used as salads.
3. Marjoram	*Marjorana hotness* (Hench) syn. *Origanum majorana* Fa: Labiatae	Marjoram is perennial while Oregano is annual or biennial but is warm temperate herbs, propagated through seed and stem cuttings respectively. Cut the entire plant before flowering, used in culinary, purposes in stews,
4. Oregano	*Origanum vulgare* (L) Fa: Labiatae	soups and sauces, yield 0.2 to 0.8% essential oil which is used in flavouring various foods.
5. Common-Mint Pepper Mint Spearmint	*Mentha arvensis* (L) *M.piperita* (L) *M.spicata*(*L*) Fa: Labiatae	Hardy perennials, 'temperate and tropical, propagation through divisions of stolons, fresh and dried leaves are used in food, medicine and cosmetics.
6. Parsley	*Petroselinum crispum* (Mill) Fa: Apiaceae	Hardy annual, subtropical, seed propagated, harvest the leaves when fresh, dried or fresh leaves, dried roots and seeds are used in flavouring dishes, medicines.
7. Rosemary	*Rosemarinus officianalis* (L) Fa: Labiatae	Perennial, temperate shrub, propagated through rooted cuttings or seeds. Both leaves and flowers are used for seasoning dishes, essential oil (0.5 to 1.5%) used in the preparation of cosmetics, soaps etc.

Continued...

8. Sage	*Salvia officinalis* (L) Fa: Labiatae	A bushy perennial, propagation-stem cuttings, harvest top growth before flowering, used for favouring sausages, canned meats, Yield 2.5% essential oil, used in mouth wash and tooth paste.
9. Tarragon	*Artemisia dracunculus* (L) Fa: Compositae	Hardy perennial herb, propagation stem cutting or root division, harvest young leaves and shoots before flower buds develop, flavour is delicious and aromatic. Yield 0.3 to 1.3% of oil used in perfumery and for flavouring.
10. Thyme	*Thymus vulgaris* Fa: Labiatae	Perennial herbaceous shurb, propagated through cuttings, harvest the leaves in bloom, aromatic and pungent flavour, yield 1% essential oil, used in food products and perfumery.
11. Savory	*Satureia hortensis* (L) Fa: Labiatae	A perennial herb aromatic, spicy quality.

CHAPTER

6

Other Spices

6.1. GARLIC (*Allium sativum* L.)

Family: **Alliaceae**

Garlic, a native of Southern Europe is one of the important bulb crops grown and used as a spice or condiment throughout India. Globally it is grown in about 12.26 lakhs ha with an annual production of 164.17 lakhs tonnes; China being the largest producer (77 %) followed by India (4 %). In India it is grown in about 1.47 laths ha annually producing about 6.45 lakhs tonnes. Europe records the world highest productivity of 21.58 t/ha as against 4.39 t/ha in India. Although it is grown in all states in India, Madya Pradesh and Gujarat have more under garlic than other states. India exports about 17,000 tonnes of garlic annually fetching a foreign exchange of Rs. 6977 lakhs.

It possesses a high nutritive value; its preparations are administered as a cure against stomach disease, sore eyes and ear ache. It is commonly used in the preparation of various dishes. Allicin, the principle of garlic, has antibacterial properties. It is a powerful drug against amoebic dysentery and is also having many other medicinal properties.

Garlic produces a group of small bulbs called cloves covered with a thin skin. The seed stalk bears both seeds and bulblets in the same head. Seed, however, is seldom used for propagation as the cloves are more commonly used.

There is no distinct variety of garlic. Local varieties are either white in colour and have fairly big bulbs with a better keeping quality and a higher yield or red in colour with pungency. The improved varieties are as below:

S. No	Name of the cultivar	Breeding method	Developed by	Yield (t/ha)	Brief Description
1.	Agri found white (G-41)	Mass selection from Bihar collection	Nashik, Maharashtra.	130	Bulb compact, silvery white with creamy flesh, bigger elongated cloves, 20-25 in number, duration 150-160 days, susceptible to purple blotch and blight, dry matter 42.78% and good storer.
2.	Yamuna Safed – 1 (G-1)	Mass selection from local collection obtained from Delhi.	Nashik, Maharashtra.	150/175	Compact bulb, creamy flesh, sickle shaped cloves, 25-30 innumbers, 150-160 days duration. Tolerant to insect pests and purple blotch, blight.dry matter 39.5%
3.	Yamuna Safed – 2 (G-50)	Mass selection from local collection in Haryana.	Nashik, Maharashtra.	150-200	Bulbs compact, attractive white creamy flesh, 35-40 cloves, duration 165-170 days.
4.	Yamuna Safed – 3 (G-282)	Mass selection from germplasm collection from Tamil Nadu	Nashik, Maharashtra.	175-200	Bulbs are creamy white and bigger sized (5-6 cm), 15-18 cloves, Duration 140-150 days.
5.	Agrifound Parvati (G-313)	Selection from exotic collection from Hong kong	Nashik, Maharashtra	175-225	Bigger sized bulbs, creamish white colour with pinkish tinge. 10-16 bigger cloves.

					Duration 250-270 days. This variety is suitable for export, developed for long day condition.
6.	Pant Lohit -1	Farmers field primitive cultivar	GB Pant University, Pant nagar, Uttranchal	-	Bulbs attractive, purple skinned variety, heavy seed potential. Tolerant to purple blotch disease.
7.	Ooty-1	Clonal selection from germplasm	TNAU, Coimbatore	150-170	Attractive white in colour. Big bulbs, 20-25 clove resistant to thrips.

There is a need to develop big cloved garlic varieties which can respond to short winter conditions of peninsular India.

Climate and Soil

It is grown under a wide range of climatic conditions. However, it cannot stand too hot or too cold weather. It prefers moderate temperature in summer as well as in winter. Short days are very favorable for the formation of bulbs. It can be grown well at elevations of 1000 to 1300 m above MSL. Garlic requires well drained loamy soils, rich in humus, with fairly good content of potash. The crop raised on sandy or loose soil does not keep for long and the bulbs too are highly lighter in weight. In heavy soils, the bulbs produced are deformed and during harvesting, many bulbs are broken and bruised and so they do not keep well in storage.

Seed Material

Garlic is propagated by cloves. All the cloves are planted except the long slender ones in the centre of the bulb. Bulbs with side growth should be discarded. Healthy cloves or bulbils free from disease and injuries should be used for sowing and about 150 to 200 kg cloves are required to plant one hectare. They are sown by dibbling or furrow planting.

1. **Dibbling:** The field is divided into small plots convenient for irrigation. Cloves may be dibbled 5 to 7.5 cm deep, keeping their growing ends upwards. They are placed 7.5 cm apart from each other in rows of 15 cm apart and then they are covered with loose soil.

June-July and October -November are the normal planting seasons for garlic.

2. **Furrow planting:** The furrows are made 15 cm with hand hoe or a cotton drill. In these furrows, cloves are dropped by hand 7.5 to 10 cm apart. They are covered lightly with loose soil and a light irrigation is given.

Recently, broad bed method of cultivation with drip and sprinkler system is being practiced by progressive growers to get higher yield.

Manures and manuring: 25 tonnes of farm yard manure is applied as a basal dose along with 60 kg N and 50kg in each of P and K. Fourty five days after planting 60kg N is applied again as top dressing. Under Karnataka condition, INM with 75; 40; 40; 40 kg NPKS+ 7.5 t/ha of Vermicompost is recommended.

Irrigation: First irrigation is given after sowing and then field is irrigated every 10 to 15 days depending upon the soil moisture availability. There should not be any scarcity of moisture in the growing season; otherwise, the development of the bulbs will be affected. The last irrigation should be given 2 to 3 days before harvesting for making it easy without damaging the bulbs. In South Indian hills, they are mostly grown as a rainfed crop.

Intercultural operations: First interculture is given with hand hoe one month after sowing. Second weeding is given one month after the first weeding and hoeing. Hoeing the crop just before the formation of bulbs (about two and half months from sowing) loosens the soil and helps in the setting of bigger and well filled bulbs. The crop should not be weeded out or hoed at a later stage because this may damage the stem and impair the keeping quality.

Harvesting

Garlic is a crop of 4 ½ to 5 months duration. When the leaves start turning yellowish or brownish and show signs of drying up, the crop is ready for harvest. The plants are then pulled out or uprooted with a country plough and are tied into small bundles which are then kept in the field or in the shade for 2-3 days for curing and drying so that the bulbs become hard and their keeping quality is improved. The bulbs maybe stored by hanging them on bamboo sticks or by keeping them on dry sand on the floor in a well ventilated room on dry floor. For taking the bulbs to the market, the dried stalks are removed and bulbs are cleaned.

Well cured garlic bulbs can be kept for 1 to 11/2 months in an ordinary well ventilated room. If dust smoke is given to it, the bulbs can be stored for 8 to 10 months. They can also be stored at 32°F with 60 percent R.H. Average yield level is 6 to 8 t/ha.

Garlic exhibits certain physiological disorders such as 1) 'sprouting of bulbs in the field'- excessive soil moisture and nitrogen supply, 2) 'splitting' – delayed harvesting or irrigation after long spell of drought and 3) 'rubberization' - controlled by application of micro nutrients $ZnSO_4$ and ammonium molybdate or neem cake application and growth regulator like GA.

Plant Protection

Thrips cause withering of the leaves. Application of chlorantraniliprol 25 18.5 % SC(0.25 ml/L) or dimethoate 30 EC 1 ml/L will check the incidence.

Leaf spot is the most important disease. Spraying Dithane M.45 at fortnightly intervals at 2.5 g in one litre of water is recommended.

6.2. VANILLA (*Vanilla planifolia* ANDR.)

Family: **Orchidaceae**

Vanilila is native of the Atlantic coast from Mexico to Brazil. It is grown on a plantation scale in Java, Mauritius, Madagascar, Tahiti, Seycheles, Zanzibar, Brazil and Jamaica and other islands of the West Indies. Malagassy Republic grows 70 to 80 percent of the world's crop of Vanilla bean followed by Reunion. The world production is estimated to be 2000 to 3000 tonnes per annum. U.S.A. is the largest importer. This spice was introduced to India as early as 1835. Its commercial cultivation is now restricted to Wynad of Kerala and Nilgiris of Tamil Nadu. Recently, the demand for natural vanilla is on the higher side.

Botany

It is an orchid, belonging to the family orchidaceae. There are two important species of vanilla viz., **V. planifolia** and **V. pompana.** The former species produces short thick pods whereas the latter one has the largest pods. **V. planifolia** has opposite, sessile leaves of 10 to 23 cm long which are oblong in shape. Aerial roots are seen on nodes by means of which the plant clings and climbs. Flower is borne in leaf axils. The flowers have three sepals and three petals which are green in colour. There is a central column of which the stamen and pistil are united with one of the petals modified to form the lip like structure called labellum. The column has to its apex two pollen sacs covered by a hood-like structure, the anther cap and the stigma is prevented from coming in contact with the anther by flap like projection known as rostellum (Fig. 11). Natural pollination is therefore impossible. In its native home, melapone bees and humming birds aid in pollination.

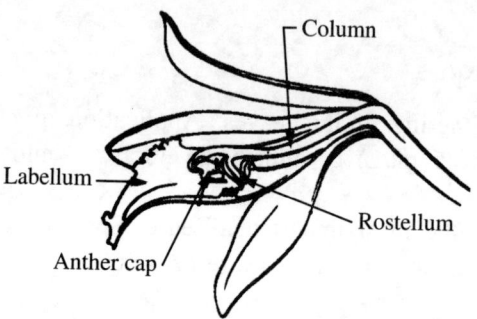

Figure 11   Structure of a vanilla flower

Climate and Soil

It prefers humid moist tropical climate having a temperature range of 25-32°C, It grows well at an elevation of 700 to 1500 metres with an annual rainfall of 250 cm well distributed for a period of nine months and dry period for three months. It is adaptable to a wide range of soil types provided there is plenty of organic matter i.e., humus and proper drainage without any water stagnation. It prefers a pH range of 6 to 6.5.

Propagation

Vanilla is generally propagated by stem cuttings. Vines of 60-120 cm long are selected as planting material. The vines are coiled and buried inside the soil. Plants raised from lengthy cuttings commence early flowering whereas the plants raised from shorter cuttings take 3 to 4 years forbearing. Therefore, cuttings with less than 5 to 6 internodes and 60 cm in length should not be used for planting. As the planting materials are a limiting factor, recently tissue cultured planting materials are made available by some private companies and Spices Board.

Establishing the Plantation

Vanilla requires support for growing. It flourishes well in partial shade. Vines may be trained on trellises on trees having low branching with rough bark and small leaves. Trees like *Jatropha, Plumeria alba, Casuarina equisetifolia, Erythrina, Glyricidia, Bauhinia* or Silver oak are now used for this purpose.

The supports are planted at a spacing of 2.5 to 3.0 m between rows and 2 m within the row making a population of 1600 to 2000 per hectare. If limb cuttings are used for planting, they should have roughly 4 to 6 cm diameter and about 1.5 to 2 m length. The supporting saplings may be established 6 months before planting vanilla cuttings. Vanilla is generally planted at a time when there is a slight wetting weather.

While planting the cuttings, three to four basal leaves in the cutting should be removed and this defoliated portion is laid on the loose soil and covered with a thin layer of about 2 to 3 cm soil. The growing end is gently tied to the support for clamping by the aerial roots. Artificial shade with any suitable material may be provided to the cuttings. It makes four to eight weeks for the cuttings to strike roots and to show initial signs of growth.

Maintenance: Once established, the vines have to be given constant attention. Any operations done in the plantation should not disturb the roots which are mainly confined to the mulch and surface layer of the soil. When the support trees grow up, they are pruned early to induce branching so as to give more shade and protection to the growing vines. If the trees are evergreen types, they are pruned before the commencement of heavy rain to allow in more sunlight. The pruned vegetation is chopped and applied as mulch in the plantation.

Manuring: The decomposed mulch is the main source of nutrients to vanilla. Animal sources of manure are not generally applied. Annually, vine may be fed with 40 to 60 g N, 20 to 30 g P_2O_5 and 60 to 100 g of K_2O. The above quantity may be given in two or three splits for efficient uptake. Part of the above fertilizers may be also given through foliar spray since they respond well to it.

Training: If the vine is permitted to grow up on a tree, it will rarely blossom, so long as it is growing upward. Hence, vines are allowed to grow up to 1.50 m and then trained horizontally on the branch of supports and later coiled round them. This induces more flower production in this portion of the vine.

Flowering: The vines commence flowering in the second or third year depending on the length of cuttings used. Due to the peculiar structure of the flower described earlier, artificial pollination by hand is the rule for fruit setting. The procedure involved is simple and done easily by children and women. Using a pointed bamboo splinter or pin, anther is pressed against the stigma with the help of thumb and thus smearing the pollen over it. Generally, 85 to 100 per cent success is obtained by hand pollination. The ideal time for pollination is between 6 am to 1 a.m. unfertilized flowers fall off within two or three days. Normally 5 to 6 flowers per inflorescence and a total of not more than 10 to 12 inflorescence per vine are pollinated. The excess flower buds are nipped off to permit the development of other pods. Pods take six weeks to attain full size from fertilization but takes 4 to 10 months to reach full maturity depending upon the locations.

Harvesting and Curing

When immature, the bean is dark green in colour, but when ripe yellowing commences from its distal end. This is the optimum time for harvesting the bean. If left on the vine the bean turns yellow on the remaining portion and starts splitting, giving out a small quantity of oil reddish brown in colour, called the Balsam of Vanilla. Eventually they become dry, brittle and finally become scentless. Therefore, the artificial methods are employed to cure vanilla. Vanillin is developed as a result of the enzyme (β- glycosidase) action on the glucosides contained in the beans during the process of curing. Basically any curing method involves the following four stages.

1. Killing the vegetative life of the beans to allow the onset of enzymatic reaction.

2. Raising temperature to promote this action and to achieve rapid drying to prevent harmful fermentation.

3. Slower drying for the development of different fragrant substances.

4. Conditioning the product by storing for a few months. The following are some of the methods:

(i) **Peruvian process:** Curing is done by hot water. In this process the pods are dipped in boiling water. The ends are tied and hung in the open. They are allowed to dry for twenty days. Later they are coated with castor oil and afterwards tied up in bundles.

(ii) **Guiana process:** The pods are collected and dried in sun till they shrivel. Later they are wiped and rubbed with olive oil. The ends are tied up to prevent splitting and then bundled.

(iii) **Mexican process:** The harvested pods are kept under shade till they shrivel. Then they are subjected to sweating. This operation is carried out for two days depending on the weather conditions. In warm weather, the pods are spread over blankets and exposed to the sun. During midday the blanket is covered over and bundled and left in the open for the rest of the day. They are wrapped in blankets in the night to maintain continuous fermentation and sweating. The pods should be wrapped in blankets when they are hot to touch. This process is repeated for 7 to 12 days till they become dark brown in colour, soft and flexible. They are packed in tins and sealed. The Mexican process yields 4.15 to 4.40 per cent of vanillin content.

When the weather is cloudy, the pods are bundled in bales and wrapped with woollen cloth covered with banana leaves. They are subject to radiation of heat by maintaining the temperature of air-oven at 50°C for 24 hours. Thereafter, they are dried to change the colour. Then they are spread in dry place and finally packed and sent to the market.

The most desirable beans will be 18 to 25 cm long, dark brown, highly aromatic, fleshy, free from mold, insects, and blemishes and some what oily in appearance. There are three grades viz., grade-1, which includes whole beans of minimum 11 cm length, and grade 2 and 3 will have a minimum of 8 cm length.

Vanillin extract is taken from the cured beans by hydroalcoholic extraction. The vanillin content of the properly cured beans will not be less than 2.5 percent.

Yield

The yield of vanilla varies depending upon the age of vines and the method of cultivation. Normally it starts yielding from the third year and the yield goes on increasing till the seventh or eighth years. Thereafter, it slowly declines till the vines are replanted after 14 to 15 years. Under reasonable level of management a middle aged plantation may yield 300-400 kg cured beans per hectare.

Plant Protection

Anthrancnose *(Calospora vanillae)* is the most serious disease. It attacks almost all parts. Root rot, *Fusarium betatis* var. *vanillae* is a limiting factor in certain areas.

The bug, *Trioza litseae* is the major pest attacking the buds and flowers of vanilla and it can be managed by spraying dichlorvas (1 ml/L).

6.3. SAFFRON (*Crocus sativus* L.)

Family: **Iridaceae**

Saffron *(Crocus sativus)* is one of the ancient spices known to man and it has its own royal position among all other spices. It is grown mainly in Kashmir in an area of 3517 ha producing about 70.89 quintals with a productivity of 1.80-2.50 kg/ha as against a very higher productivity of 6.80 kg/ha in Italy. Nearly 50 % of Indian saffron is exported. It is as precious as gold (often called as 'orange gold') not only because of its low production and high demand, but also because of its wide range usage in medicine and religious rituals. It is used to colour foods even at 1 ppm concentration which gives distinct yellow tinge. It has 0.5 to 1.0 percent essential oil. The main principle is Cicrocrocin.

It is a shorter perennial herb with a globular underground corm and six or more radial, narrow and linear leaves. The pungent smelling bluish-violet flowers are borne singly and their strikingly dark red or orange tripartite funnel shaped stigmas are used as the spice. In India, it is cultivated only in Kashmir. Bulbs are planted in August on flat ground having good drainage. Relatively, low annual rainfall of about 35 to 45 cm is desirable. Heavy rains at bloom time or early autumnal frost is not desirable.

The bulbs continue the cycle for 10 to 15 years. Initially, the growth is rather sluggish for the first two years and reaches a peak by the third year and after six years; the bulb is uprooted for subsequent planting, leaving the ground fallow for nearly 4 to 5 years.

The flowering season lasts for nearly two weeks. By the end of October, the saffron usually flowers. At this time pickers complete picking of flowers early in the morning before sun gets too hot or before the flowers wilt on their stem. Later, the golden coloured styles from the each of the flower are broken off just below the stigma and are dried in the sun. Drying percentage is about 20%. This finest, purest and most expensive saffron is known as Shahi saffron in trade and the rest of the flower parts go to make up Mogra saffron and Lache saffron the inferior grades. About 1.5 million flowers on drying give one kg of this spice. On an average, from one hectare about 160 kg of fresh flowers i.e., about 5kg of dried saffrons can be obtained.

6.4. PAPRIKA (*Capsicum annuum* L.)

Family: **Solanaceae**

Paprika refers to the bright red, sweet to mildly pungent fruits of ***Capsicum annuum.*** These fruits differ in size and shape, some are round, some conical, some long and pointed. They are quite fleshy and very mild. The spice is derived by drying and grinding the deseeded and destemmed ripe fruits which yield a sweet or mildly pungent red coloured powder, rich in vitamin C. It is mainly used for adding colour to the finished product and to make the product more acceptable and pleasing to eye. The powder is used to flavour and garnish foods like cheese, eggs, fish dishes, salads, sauce and rice. The principle colouring matter is the carotenoid pigment, called capsanthin. The outer fleshy pericarp carries the major portion of the colouring matter while the inner tissues and seeds contain the pungent chemical capsaicin.

Paprika powder and its oleoresin are in great demand in the Western World. Spain, Hungary, Morocco, Yugoslavia and USA are the chief producers of the paprika. Though it is not cultivated in India now, there is a good scope to grow this crop in the hill stations of South India and North India and also in plains of North and Central India where mild climate prevails.

Cultural practices for paprika are as good as for chillies. The seedlings are raised in nurseries. About one kg of seeds raised in 200 sq. m nursery area will be adequate to supply seedlings for one hectare. 6 to 8 weeks old seedlings may be transplanted at a spacing of 1m × 0.3 m in ridges and furrows. First harvest will be in about 80 days from transplanting. Fruits are harvested as they reach maturity. About 5 to 6 pickings may be necessary. The harvested ripe fruits are thus cured and dried. Average yield will be around 2000 kg per hectare.

6.5. ASAFOETIDA (*Ferula foetida* Regel)

Family: **Apiaceae**

Asafoetida is the oleogum obtained from plants such as *Ferula alliacea* Boiss, *F. foetida* Regel and *F. galbani* Boiss. This plant is found in the mountain slopes of Afghanistan, Iran, Turkey and Kashmir. Asafoetida is used as a spice, flavouring compound and in pharmaceutical preparations. India is importing about 400 tonnes of this spice at a cost of Rs. 200 lakhs annually.

It is a perennial plant, reaches a height of 1 to 1.5 m and have massive carrot shaped roots. These rhizomes produce two types of plants; one is a male plant producing inflorescence and the other female plant which has only foliage and no inflorescence. The female plants produce an exudation in the form of a very thick and sticky paste sap from the underground rhizome, called as asafoetida.

During the onset of spring, sprouts come out from the rhizomes of the female plant and putforth foliage. It takes nearly one month for the green foliage to turn yellow. It is the right stage for tapping the rhizomes for asafoetida.

The tapping is done in stages. Initially, the yellow foliage is removed and then the top of the rhizome and the brush like mass is covered with loose earth and gravel and is left for 5 days. After about 5 days, the brushy mass on the top of the rhizome is pulled out and the earth and gravel around the rhizome are cleaned, exposing the top of the rhizome.

With a sharp knife, a small portion of the top measuring about 6 to 7 cm^2 is scraped out. After scraping, a sort of cover or shade is built by placing three or four stones along with border of the exposed rhizome and one stone on the top.

After two or three days of the scrapping of the top of the rhizome, the first collection of the sap is made. After collecting the sap or the paste from the first scrap, a slightly deeper scrap, say about 0.5 cm is again made on the rhizome at the same spot below the first scrap. The scraping can be repeated up to ten to fifteen times, each time a scrap going deeper into the rhizome till finally, no further sap is exuding out from the rhizome. Each time the top should be protected as stated earlier.

The sap is usually collected in pits of 1 m × 1 m × 20 cm having the sides plastered with mud and the top covered with stalks of male asafoetida plant leaving a small hole of 30 × 30 cm to allow pouring of the sap down to the pit, In the pit, the asafoetida which is initially in the form of very thicky paste matures further. The major constituents of asafoetida are resin (40 to 60%), gum (25%), and volatile oil (10-17%).

7

Value Added Spice Products

The term value added products in general indicates that for the same volume of a primary product, a high price is realised by means of processing, packaging, upgrading the quality or other such methods.

The following are the different kinds of value added products in spices.

1. **Spice oil:** It is obtained by steam distillation. The essential oil thus obtained is endowed with the major part of the flavour and fragrance properties of the spices.

2. **Spice oleoresin:** It is obtained by extraction of the dry ground spices with an organic solvent or solvent mixture such as ethylene chloride, acetone, hexane or alcohol. It contains all the volatile as well as non-volatile constituent of the spice, closely representing the total flavour of the fresh spice in the concentrated form. The residual solvent in the oleoresin should be below 30 ppm.

3. **Ground spices:** The spices are milled to the degree of fineness such as cracked, course grind, fine grind etc. as required by the user. There is a considerable heat evolved during the grinding operation resulting in flavour loss or modification. To overcome this, spices are now milled at low temperature by feeding spices and liquid nitrogen simultaneously into the milling zone. The cryomilled spices have better retention of aroma, colour and less loss of moisture.

4. **Curry powder:** It refers to the powdered blend of a number of spices. It has got a good export potential.

5. **Consumer packed spices:** The spices exported to the developed countries are consumed in three main segments viz., industrial, retail and institutional. Different packaging media are used according to the consumer preference for those parts of spices which are sold by retail. The popular packaging media are glass bottles, rigid plastic containers, metal containers and flexible pouches. The prices of such retail spice packs are higher between 500-1500 percent when compared to the prices of bulk.

6. **Improved processing technique:** Dehydrated, frozen or freeze dried green pepper and canned pepper have much scope in selected West European countries.

7. **Encapsulation of spice oleoresins:** When milled spices are stored, their essential oils evaporate over time. Micro-encapsulation preserves these valuable flavours. In this process, the milled particles are microencapsulated and then sterilized at a mild temperature to reduce the germ count. The process is suitable for a variety of spices and takes just a few seconds. Microencapsulated spices currently are making an outstanding contribution to targeted flavour development in end products, particularly when it comes to special manufacturing processes in the food industry. It has many advantages such as standardized, high flavour content, very good storage stability due to encapsulation of the valuable spice constituents, consistent flavour profile, reduced germ count without ETO treatment or irradiation and reduced enzyme activity.

These value added spice products are used in

(a) **Food industries:** processed meat, sauces, soups, cheese and dairy products, confectionery and alcoholic and non-alcoholic beverages and

(b) **Non-food industries:** cosmetics, pharmaceuticals, hygiene products and perfumery industries

Management of Mycotoxins in Spice Products

Presence of mycotoxins is not only related to the effect they might have on consumer health, but may also have an impact on world trade as mycotoxins are considered as the main hazard in border rejection notifications in the European Union. Mycotoxins are produced by a number of fungal genera primarily *Aspergillus, Penicillium, Alternaria, Fusairum, and Claviceps.*

The common mycotoxins are aflatoxins, deoxynivalenol (DON), fumonisins, ochratoxin A, patulin and zearalenone, of which the rejection due to aflotoxin is the most common one.

This microbial contamination in spices is mainly due to high moisture content due to improper post harvest handling. It is necessary that the spices should be dried to a safe moisture level of 10 % and stored at this moisture level. In actual practice it is found that though spices are initially dried properly, they tend to absorb moisture from the atmosphere due to its hygroscopic nature and also due to the prevalence of high atmospheric humidity during monsoon months. To avoid these problems, suitable packaging materials like Epoxy lined steel drums, High Density Polythene containers, PET (Poly Ethylene tetrapthalate) bottles, HMHDPE (High Molecular weight High Density Poly Ethylene) containers are to be used.

Significance of Value Added Spice Products

Annually 7600 tonnes of spice oil and oleoresins worth of Rs.910 crores are exported from India. Besides, 15250 tonnes of curry powder worth of nearly Rs.210 crores are exported from India every year and the demand for these vaue added spice products are on increasing trend every year. Spice value added products include mint products, curry powder, spice powders, blends and seasonings .The export of processed spices such as curry powder, mint products and spice oils and oleoresins accounted for 44.2 per cent of total exports and among them mint products alone account for a large volume of spice (29.7 per cent). There is a good demand for these spices oils and oleoresins in countries like USA, U.K., Canada and Germany .In these countries; they prefer to use more of spice oils and oleoresins instead of the spices as such because of certain advantages given below:

1. The oils can be easily incorporated into liquid flavour concentrate or oils and fats and the flavour strength can be controlled.
2. They do not impart any colour to the product.
3. They are free from microbial, insect and rodent contamination.
4. They require less storage space and keep stable over a long period.
5. More volumes can be handled per unit area
6. Flavour value is uniform.

There are 15 spice oleoresin manufacturing units in India today and most of them are located in and around the city of Cochin in the state of Kerala, which has a long history as a major export centre for spices from the Malabar Coast. The concentration of these manufacturing activities in

Cochin is primarily due to its proximity to India's major spice growing regions and its connectivity by sea and air. These transport facilities are in the process of further improvement: there is already a new international airport and Cochin will get an international container transhipment terminal soon. This will certainly give this city the status of an outsourcing destination, not only for the flavour and fragrance industry, but others as well.

Agencies Supporting Spice Industry

1. **Spices Board**

   Spices Board, India, under the Ministry of Commerce and Industry, is the apex organization for promoting Indian spices and spices products worldwide. The Board was constituted on 26th February, 1987 as per Spices Board Act, 1986 by merging the erstwhile Cardamom Board and Spices Export Promotion Council. The mandate is mainly export development and promotion of 52 scheduled spices and their processed forms.

2. **World Spice Organization (WSO)**

   Another development is the formation of WSO which is a common platform for all the stake holders of the spice industry viz farmers, processors, academia, and end-users to come together and work for its sustainable development.

3. **Spice Parks**

   Recently, Government of India has established 10 spice parks for major spices in the hot spice growing area which function as a fulcrum for development of spice industries. The basic concept of such park is to provide common infrastructure facilities for both post harvest and processing operations which also aims to backward integration by providing rural employment. These parks have processing facilities at par with international standard in which the produce could undergo cleaning, grading, grinding and packing etc. It also provides the educative service to all the farming and trading communities about good agricultural practices to obtain quality standards prescribed by the importing countries.

4. **Directorate of Arecanut and Spices Development , Calicut**

   This directorate functioning under Ministry of Agriculture, GOI is responsible for coordinating and monitoring the activities on development of spices including aromatic plants. It supplements the developmental efforts of the State Govt. by making available the nucleus planting materials of various HYV of spices and aromatic plants across the country, through a centralised nursery programme at major centres of cultivation. It also disseminates periodically the seasonal crop prospects, area coverage, price trend etc. for the use in the planning process.

PART-II
PLANTATION CROPS

8

Introduction

The term *plantation crops* refer to those crops which are cultivated on an extensive scale in a large contiguous area, owned and managed by an individual or a company. The crops include tea, coffee, rubber, cocoa, coconut, areca nut, oil palm, palmyrah, cashew, cinchona etc. These plantation crops are high value commercial crops of greater economic importance and play a vital role in our Indian economy. The main drawback with this sector of crops in India is that major portion of the area is of small holdings (except Tea) which hinder the adoption of intensive cultivation. In the case of coffee, 97.13 percent of the growers have holdings below ten hectares and in rubber, 82 percent of the total area is of small holdings having an average size of 0.5 ha. The economic importance of these crops is

1. They contribute to national economy by way of export earnings. These crops occupy less than 2 percent of the total cultivated area (i.e. 1.82 percent of total cropland) but they generate an income of around Rs. 4450 crores. Its share in total export earnings from India declined from 12.70 % in 1970 to 0.68 % in the recent time.

2. India is the leading country in the total production of certain plantation crops in the globe. For instance, our production meets the share of 47 percent in tea and 66 percent in each of cashew and areca nut.

3. Plantation industry provides direct as well as indirect employment to many millions of people. For instance, tea industry offers direct

employment to 10 lakhs and indirect employment to 10 lakhs people, while cashew processing factories alone provide employment to 3 lakhs people besides 2 lakhs farmers are employed in cashew cultivation.

4. This sector of late experiences much problem because of non availability of laboureres. It is reported that in the last 10 years 60 percent workers have migrated off the tea plantations and similar situations also prevail in other plantation crops also. This forces this industry to move for mechanization.

5. Due to globalization, this industry is undergoing lot of changes and is also subjected to the laws of demand and supply leading to wide price fluctuation.

6. Plantation industry supports many bye-product industries and also many rural industries. For example, coconut husk is used to produce coir fibre annually to a tune of 2,19,600 tonnes in India.

7. Plantation sector preserves the environment. These crops help to conserve the soil and ecosystem. Tea planted in hill slopes and cashew in barren and waste lands protect the land from soil erosion during the rainy season or due to heavy winds.

8. Government of India is attaching much importance to this sector through various commodity boards such as Tea Board, Coconut Board, Coffee Board and Rubber Board besides Directorate of Cocoa and Cashew development and Directorate of Arecanut and Spices Development to coordinate the research and development programme in these crops.

9

Tea

TEA (*Camellia sp.*)

Family: **Camelliaceae**

Tea, the oldest known beverage, is native of China in South East Asia. It was known to the Chinese as early as 2737 BC, but attained the status of a popular drink in England in 1664 A.D. It was planted on a large scale in North India in 1834 while in South India from 1859 to 1897 in different tea growing districts.

As on today, 38 countries grow tea and among which India, China, Sri Lanka and Indonesia have major share in area and production. Globally we are producing 4100 milion kg and India has been pushed to 2nd position by China for the first time in 110 years in 2006. China is the major producer (1475 million kg) followed by India which is the largest consumer and exporter in tea industry. The area and production of major tea growing states of India is as follows:-

State	Area in ha.	Production (M.Kg.)	Production (%)
Assam	322214	508741	51.47
WestBengal	115095	226362	22.90

Other states in North India	22304	12344	2.26
Tamil Nadu	80462	167229	16.92
Kerala	37137	68347	6.92
Karnataka	2141	5305	0.57
Total	579353	988328	100

Tea sector in India is largely organized since 72 % of the total area is from 1686 estates while the balance (28 %) is from small growers .India annually exports about 200 M.Kg. of tea (33%) fetching Rs. 715 crores as foreign exchange which accounts for 7 percent of total foreign exchange. It is a labour intensive crop providing direct employment to 10 lakhs and also indirect employment to 10 lakhs people. India's average productivity is 1690 kg /ha and the cost of production of tea remains the highest among all the tea producing countries. Consumption of tea varies with countries; 1 kg per head in Sri Lanka and Pakistan while it is 2 kg per head in Ireland and UK and in India it is only 800 g per head, yet it is the largest consumer because of the population.

Botany

Tea belongs to the genus *Camellia* and family Camelliaceae. The original species which produce tea were *C. assamica* (Assam jats), *C. sinensis* (China jats) and their natural hybrid, *C. assamica* subspecies *lasiocalyx* (Indo China or cambod type.). Being a highly cross pollinated crop, the present day seedling populations are mixture of both the above two species, however, from their major share of characters-Assam or China type can be distinguished by the following characters:

	Assam	China
1.	It is a tree	It is a shrub
2.	Few robust branches.	Branches abundant and whippy.
3.	Large, glossy leaves.	Small leathery leaves.
4.	Light to medium green	Dark green colour.
5.	High yield and medium quality.	Low yield but good quality.
6.	Susceptible to drought and frost.	Hardy and resistant.
7.	Sparse flowering.	Profuse flowering.

Morphologically, tea is an evergreen shrub or tree, leaves are simple, alternate, serrate, flower bisexual, with superior ovary, fruit is a capsule.

Varieties

Clonal selection from seedling population was taken up by UPASI, Tea Scientific Department, Cinchona and also by other Tea Research Institutes. This has resulted in the release of a number of clones. UPASI has so far released 27 clones. Certain outstanding clones released by other Institutes are also used in South India. The important features of the promising clones are alone presented below:

Clone	Remarks
UPASI-2 (Jayaram)	An average yielding clone, suitable for all elevations, tolerant to drought and wind.
UPASI-3 (Sundaram)	Very high yielding and quality clone.
UPASI-6 (Brooklands)	Fares well at mid and higher elevations.
UPASI-8 (Golconda)	High yielding, suitable for all elevations.
UPASI-9 (Athrey)	High yielding, fairly tolerant to drought, can withstand slightly high pH.
UPASI-10 (Pandian)	Hardy clone, resistant to drought and wind; suitable for high elevation.
UPASI-14 (Singara)	Quality clone, suitable for higher elevations.
UPASI-17(Swarna)	A high yielding clone.
UPASI -20	Hardy clone
UPASI-24	Yield
UPASI-25	Hardy clone
UPASI-26	Hardy
UPASI-27	Yield
UPASI-28 (BSB-1)	Hardy &Yield
TRF-1	UPASI–21 X TRI-2025
TRF-2	China seedling –unknown parents
TRF-3	UPASI–10 X TRI-2025
TRF-4	CR-6017 X UPASI - 8
TRF-5	UPASI–21 X TRI-2025
TRI-2024 (Sri Lanka)	High yielding clone.
TRI-2025 (Sri Lanka)	Average yielding; hardy clone.

The 'bicloanal seed stock' programme is to combine the desirable attributes found in different released clones through hybridisation, to evolve cultivars that are high yielding with quality and tolerance to drought, diseases and pests. The performance of the biclonal progenies of the combinations given below has been found to be promising for commercial planting under South Indian conditions:

UPASI Biclonal Seed Stocks (BSS)

Seedstock	Female parent X Male parent
UPASI : BSS-1	UPASI-10 X TRI-2025
UPASI : BSS-2	UPASI-2 X TRI-2025
UPASI : BSS-3	UPASI-9 X TRI-2025
UPASI : BSS-4	UPASI-15 X TRI-2025
UPASI : BSS-5	CR-6017 X UPASI-8

Due to the profuse tap root system, these biclonal seed stocks are recommended to be used as 'infills' and also for replanting in drought prone areas.

Climate and Soil

Tea is exacting in its climatic requirements. The temperature may vary from 16 to 32°C and annual rainfall should be 125 to 150 cm which is well distributed over 8-9 months in a year. The atmospheric humidity should be always around 80% during most of the time. Very dry atmosphere is not congenial for tea. It is grown in plains in North Eastern States but in South India, it is grown in hill ranges from 600 to 2200 m above M.S.L. Tea is a calcifuge crop requiring comparatively low amounts of calcium but high quantities of potassium and silicon. They can be grown in lateritic, alluvial and peaty soils. Optimum pH range is 4.5 to 5.0 and soil depth should be 1.0 to 1.5m.

Propagation

Tea can be propagated by seed and by cuttings. Seed propgation is seldom practised now. Seeds collected from the fruits of seed baries are soaked in water and only heavy seeds which sink are alone used for sowing in beds. Germination occurs in 20 to 30 days. At that stage they are carefully lifted and transplanted in polythene sleeves. They will be ready for planting in 9 months.

Vegetative propagation: The site for the nursery can be selected in a flat land or gentle slope, near to a perennial water source and easily accessible by road. It should have a good drainage and should be protected from wind, frost and wild animals etc. Approximately, 0.15 ha nursery area is required to produce 1.25 lakhs cuttings. Nursery area is to be provided with overhead shade by erecting concrete or stone pillars at a spacing of 3 × 3 m and spread with 6 mm² mesh double strand coir mat which provides about 67 percent shade.

Figure 12   Single node cutting in Tea

The cuttings for rooting are collected from mother bushes which are well maintained near the nursery area. Such mother bushes are pruned well in advance to induce juvenile shoots. These juvenile shoots are collected in the morning hours and 3 cm long cutting each with a healthy mother leaf and an active axillary bud is prepared (Figure 12). Cuttings from top tender and bottom brown wood should be avoided. These cuttings are planted in polythene sleeves (30 cm × 10 cm × 150 gauge), filled with growing medium (Jungle soil: river sand 3: 1) in the bottom and rooting medium (Red/sub soil: sand: 1:1) in the top 8-10 cm. The soil used for rooting media should have an optimum pH range of 4.8 to 5.0, if high, i.e. 5.1 to 5.5, or 5.6 to 6.0, it must be drenched with 1 or 2% aluminium sulphate solution respectively @ 1 litre per cubic foot of soil. This treatment should follow with drenching of twice the volume of plain water to wash excess aluminum sulphate. The cuttings are carefully planted at the centre of the sleeves in such a way that the petiole should not touch the soil and then they are watered. These sleeves are then covered with polythene sheets over the G.I. wire arches and the sides are tugged well to preserve moisture content. Callusing starts in 4-6 weeks and rooting occurs in 10 to 12 weeks. When 80 percent of the cuttings have rooted, the tents are opened in stages and the overhead shade is gradually reduced to harden the plants.

Grafting: Recently cleft grafting of single nodal cuttings of two varieties in the nursery and callusing them in the nursery to develop a composite plant has been followed to take advantage of drought tolerant clones as stocks and high yielding and quality clones as scions.

Planting

The land is cleared of the roots of the fallen trees and drains are taken at suitable intervals depending upon the slope to conserve the soil. In the olden days, up and down system of planting at 1.2 × 1.2 m are followed. Presently, contour planting either in a single hedge or double hedge system is followed.

Style	Spacing	Population/ha.
1. Up and down	1.2 × 1.2 m	6,800
2. Contour planting single hedge	1.2 × 0.75	10,800
3. Contour planting double hedge.	1.35 × 0.75 × 0.75	13,200

The last method has many advantages over the first two viz., early and high yield, better soil conservation, less weed growth in the hedge and efficient cultural practices. Planting season normally coincides with June/July and September/October for South West monsoon and North East monsoon areas. Pits of 30 × 30 × 45 cm size are dug and plants of 12-15 months old are planted by removing the polythene sleeves. Immediate after planting, plants are staked to prevent wind damage.

After care

Immediately after planting, the soil surface around the plants should be mulched, usually cut grasses of gautemala are employed for this purpose. About 25 tonnes of grass is required to mulch one hectare. Care must be taken to keep the mulch materials away from the collar region lest they may cause collar diseases. If there is a dry weater, mud tubes or etah tubes may be buried 15cm deep near the plant in a slanting position and one litre of water per plant may be poured or injected at weekly intervals. This subsoil irrigation helps to minimise the causality besides encourages developing deeper roots. Another method to overcome drought is through application of Kaoline. Suspension of 12 percent Kaolin and 0.33% vinofan in 100 litres of water is applied @ 25ml/plant .This reflects light and reduces leaf temperature and transpiration. This should be supplemented with sub-soil irrigation when first leaf droops.

Drip fertigation is also gaining popularity in tea plantations. Drip irrigation at 2 mm daily with fertigation of $N:K_2O$ 1:1 in 10 splits shows an enhancement in yield. The sprinkler irrigation at 6.25 cm/ha at 2C days interval during the dry months also helps to improve the green leaf yield.

Shade and its management: Tea requires filtered shade and if it is exposed to direct sun, its growth is affected. Shade is hence essential and beneficial to tea as

1. It regulates the temperature.
2. It minimises the effects of drought and radiation injury.
3. It increases the soil fertility (leaf litter adds about 8-10 tonnes of organic matter per ha/year).
4. It helps in recycling of nutrients.
5. It helps in getting even distribution of crop.
6. It serves as wind break.
7. It reduces the incidences of pests.
8. It generates additional income by way of timber and fuel.

The main drawbacks of shade are 1) increased incidence of blister blight 2) competition with main crop for moisture and nutrients and 3) reduced response to applied fertilizers.

In South India, Silver oak (***Grevillea robusta***) is used as the permanent shade tree as it possesses the desirable characters of a good shade tree like

1. It must be an evergreen tree, easy to propagate having quick growing and deep rooted characters.
2. It provides filtered shade and withstands frequent lopping.
3. It tolerates wind and frost.
4. It does not have allelopathic effect.
5. It has commercial timber value also.

The seeds of silver oak are very light and fresh seeds are sown in the raised beds in line at 5 × 2.5 cm spacing during December/January and covered with thin layer of sand. They germinate in 10 to 20 days. When they develop first pair of leaves, they are pricked and transplanted in the polythene sleeve and kept in shaded area. Shade is gradually removed to harden them. Six to nine months old seeds are ready for planting.Silver oak seedlings are initially planted along the tea rows at 6m × 6m spacing. As and when they grow, lower branches may be lifted periodically. When it attains 8 to 9 m height, the tree is pollarded at a site having a girth of 10 to 15 cm. Below the pollarded site, one branch in each direction may be left in 3 to 4 tiers and the excess ones are removed. Before every monsoon, the lateral branches, erect growing branches and shoot growth in the main stem are removed.When they are about 10 years old, shade trees are thinned out before pruning season by removing the alternate trees in the east-west rows. Such trees earmarked for felling should be ring barked 2 years prior to felling to have lesser starch reserves in the stump.

In mid elevation till silver oak seedlings grow, temporary shades are employed. *Indigofera teysmanii* is normally planted at 3 × 3 m spacing. They are removed once silver oak trees get established. In frost prone areas, instead of Grevilleas, another shade tree viz., seedlings of *Acacia mearnsii* are planted at 3 × 3 m spacing and are pollarded at 3 m height. Being a shallow rooted and surface feeding tree, these trees are replaced once in two pruning cycles. UPASI now recommends alternate shade trees to Grevillea robusta such as *Toona sinensis, Toona ciliata, Melia dubia* which are native to Western Ghats.

Weed control: Weeds will be a problem in young and pruned fields. Manual weeding is never recommended in tea lest more soil erosion and damage to surface roots and collar regions. Hence, the following chemical weed control is alone recommended in tea.

Type of weeds	Herbicides	Dosage
Dicots	Paraquat (gramoxone)	1.12lit./ha.
Dicots	Sodium salt of 2,4-D (Fernoxone)	1.4kg./ha.
Grasses	2,2-Dichloro propionic acid (Dalapon)	5.6 kg./ha.
	Glyphosate	2.3 lit./ha.

Training young tea: In the young tea, when it has established well, centering i.e. removing the growing point leaving 8 to 10 mature leaves from the bottom, is done to induce secondaries. When the secondaries reach more than 60 cm, they are tipped at 50-55 cm height by removing 3 to 4 leaves and bud to induce tertiaries. Therefore, plucking at mother leaf stage is continued for better frame development. It takes nearly 18 to 20 months from planting to reach regular plucking field stage.

Pruning: Pruning is done in tea (1) to maintain the convenient height for plucking (2) to induce more vegetative growth (3) to remove dead and defunct wood and (4) to remove the knots and interlaced branches. Pruning is normally done 4 to 6 years interval depending upon the altitude of the garden, nature of the tea materials etc. The bushes marked for pruning should have adequate starch reserves in roots otherwise the sprouting following pruning will be less. This can be normally tested by the common Iodine test and if the starch reserve is less, bushes are allowed to rest for 2 to 3 months. The different types of pruning are as follows:-

	Type of pruning	Pruning height (cm)	Season	Remarks
1.	Rejuvenation pruning	20-ChinaJat 30-Assam Jat	April-May	Done in old bushes affected with canker and wood rot to invigorate the new healthy branches. Not done regularly.
2.	Hard pruning	30-45	Apr.-May	First formative pruning done to a young tea.
3.	Medium pruning	45-60	Aug.-Sept.	Normal pruning where-ever frames are healthy.
4.	Light pruning	60-65	Aug.-Sept.	Normal pruning where-ever frames are healthy.
5.	Skiffing	65	Aug.-Sept.	Mainly to postpone pruning and to encourage better frame development.

Immediately after the rejuvenation or hard pruning, the cut ends are smeared with a paste made of copper oxychloride and linseed oil (1:1). The pruned material, consisting of only small twigs and leaves are buried in trenches of 30 cm width and 45 cm depth taken across the slope in alternate rows. The pruned bushes are given washing with 10% lime solution using No.IV nozzle of power sprayers in order to kill the epiphytic growth of moss and lichen so as to induce early and even bud break. Lime washing also minimises sun scorch to the bush frame.

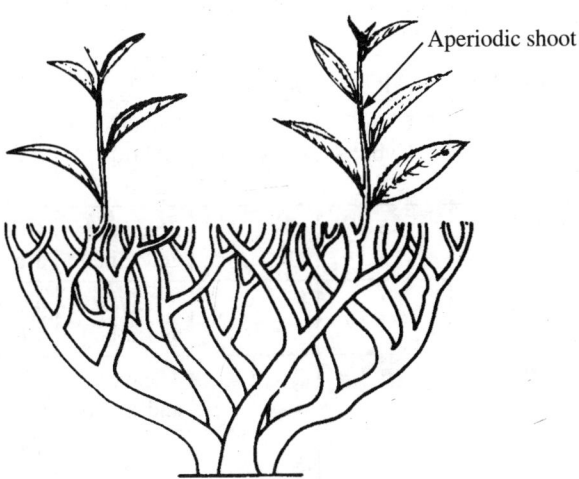

Aperiodic shoot

Figure 13 Aperiodic shoot in pruned tea bush

The buds from the pruned shoots grow in steady succession without any cessation of growth. These are known as aperiodic shoots or primary shoots (Figure 13). These primary shoots should be (Figure 14) induced to produce flush shoots, otherwise known as *periodic shoots* by regular **tipping** operation. Tipping is the removal of terminal portion of the shoot and it varies with jats and pruning height as given below. Tipping height refers to the number of leaves that must be left above the pruned cut while **tipping in material** refers to that portion of the terminal shoot which must be tipped off.

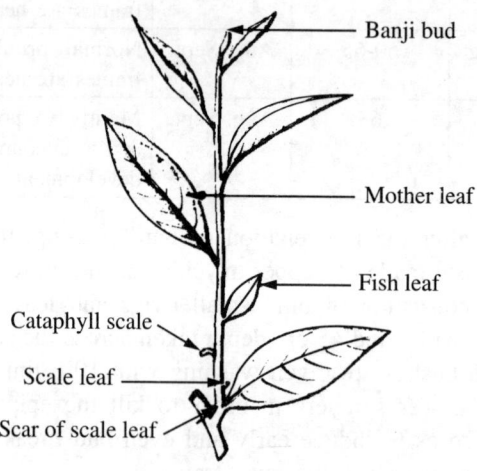

Figure 14 Periodic shoot in Tea.

Pruning height(cm)		Tipping height (cm)		Tipping in material	
China Hybrid	Assam/ Assam Hybrid	China Hybrid	Assam/ Assam Hybrid	China Hybrid	Assam/ Assam Hybrid
35-45	35–55	5	4	3 leaves and a bud	4 leaves and a bud
45-55	55–60	4	3	4 leaves and a bud	4 leaves and a bud
55–75	60–75	2	2	4 leaves and a bud	4 leaves and a bud
–	>75		1		4 leaves and a bud

Manuring

Tea responds to manuring and it has been estimated that to produce 100 kg. of made tea, tea plant utilises on an average 10.2, 3.2 and 5.4 kg of N, P_2O_5 and K_2O per ha. Manuring in tea starts from nursery stage itself. Once they strike roots (after 4 months) 30 g of soluble mixtures (Ammonium phosphate (20:20) 35 parts, Potassium sulphate and Magnesium sulphate each 15 parts and Zinc sulphate-3 parts) is dissolved in 10 litres of water and is applied with rose can for about 900 plants. This must be repeated at 15 days intervals.

Manuring to the tea in the field depends on many factors viz., young or mature tea, organic and nutrient status of the soil, soil pH, nature of pruning and yield potential of the tea bushes.

Nitrogen: The recommendation for mature tea is based mostly on soil organic matter status and anticipated yield. For a field with medium organic matter status the following rates of application is suggested for every 100 kg of made tea anticipated:

Yield level (kg/ha)	Rate of Nitrogen (for 100 kg. of made tea)	No. of split applications
Up to 2000	10 kg	4
Above 2000	5 kg	5

Twenty percent of the total nitrogen is given in the form of Ammonium sulphate during March/April. Urea is recommended in May/June and receding monsoon months avoiding very wet and very dry periods and it will come to 65% of total nitrogen. Fifteen percent of the total nitrogen is applied in the form of Calcium Ammonium Nitrate during pre-winter (November/December).

Potassium: Nitrogen and potassium are always applied together. NK ratio of 1:1 is used for plucking fields while for a pruned field 2:3 NK ratio is recommended. For rejuvenation pruned field 1:2 NK ratio is suggested. The enhanced rate of potassium application in the pruned year is to encourage formation of healthy frames. Muriate of potash is the source of potassium used in tea fields. The NK fertilizers are applied by broadcast for mature tea and is broadcast and dibbled in along the drip circle for young tea. The interval between two successive applications should be atleast 3-4 weeks.

Phosphorus: This nutrient is applied once in alternate years @ 60 -80 kg P_2O_5/ha for fields yielding less than 3500 kg/ha if soil tests P value is less than 22.5 ppm. For fields yielding above 3500 kg/ha, 80 -100 kg

P_2O_5 if soil tests P value is less than 44 ppm. The soils being acidic, rock phosphate could be advantageously used. The fertilizer should be placed at 15-22 cm depth.

UPASI recommends 20-25 kg /ha each of Azospirillum and Phosphobacterial formulations per ha to reduce the quantity of chemical fertilizers. A reduction of 15 – 20% nitrogenous fertilizers and 50% of phosphate fertilizers is possible by applying biofertilisers. These bacteria are capable of producing plant growth promoting substances, vitamins and antibiotics which will improve plant growth and protect the plants from minor root pathogens. Further application of digested coir compost @ 3-6 MT/ha along with these biofertilesrs are known to improve the green leaf yield and the soil physical, chemical and biological properties.

Micro nutrients: Among the micronutrients, zinc deficiency is often manifested in young shoots characterised by reduced leaf size, resetting, chlorosis and formation of more banji shoots. Application of Zinc sulphate @ 8 to 12 kg/ha every year is the general recommendation. The above quantity can be given in 4 to 5 split applications during the high cropping months i.e. during April/May and September/October. It has been found beneficial to combine Zinc sulphate spray with other micronutrients viz., Manganese sulphate @ 15.5 g/10 litres and Boric acid @ 5.5 g/10 litres along with Naphthalene acetic acid @ 2.2 ml/10 litres of spray volume. Addition of 2kg $MgSO_4$ along with Zinc sulphate 2 kg is also recently recommended. Ammonium polyborate can be used as an effective alternative to Boric acid @ 100 g/200 L of water per ha in micronutrient spray.

Liming: In the hill soils, due to the leaching of bases by rain and also due to the incessant application of acid forming fertilizers, the soil pH is often reduced which affects the physical and chemical properties of soil. Therefore, periodical application of lime is essential to amend the soil and maintain optimum pH. Agricultural lime (Calcium carbonate) and dolomite lime (Calcium manganese carbonate) are generally recommended for tea soils. The rate of application is based on soil pH, rainfall, fertiliser usage and length of the pruning cycle. Roughly lime @ 1.5 m/ha for a pH between 4.5 to 4.9, 3.0 m/ha for a pH between 4.0 to 4.4 and 4.0 m/ha for a pH of less than 4.0 is suggested. The lime is applied by evenly broadcasting prior to pruning once in a pruning cycle. First manuring following liming can be had after 6 weeks and a minimum of 15 cm rainfall should have been received during this period.

Plucking

Plucking consists of harvesting 2 to 3 leaves and a bud. It is the most labour intensive operation in a tea industry and also decides the yield and quality of made tea. Normally, a pluckable shoot takes 60 to 90 days for harvesting since its sprouting from the axillary buds. When the shoot is plucked up to mother leaf, it is known as light plucking and if it is plucked below mother leaf, it is called hard plucking. The plucking interval and plucking standard in relation to cropping is given below:-

Cropping pattern	Months	Plucking interval
High cropping or Rush cropping (60% of total crop)	April - June and October - December	7-10 days
Low cropping or lean cropping (40% of total crop)	July - September and January - March	12-15 days

It is essential to add one tier of active maintenance foliage to the bush every year. This is done by mother leaf plucking during January to March. During the rest of the period level plucking can be carried out. Consequent to plucking, bush height increases every year in the order of 10 cm over tipping height in the first year, 7.5 cm, 7.5 cm, 5 cm and 5 cm over the previous year height in the second, third, fourth and fifth year respectively. In some places, a scissor like mechanical/shear harvester is employed to pluck during the high cropping period. It helps to manage the high crop and overcome labour scarcity. UPASI recommends use of two men operated machines to harvest the clonal fields planted on gentle slopes of less than 15 % while one man operated machine can be used up to 30 % slopes and hand operated shears can be used for very steep terrains of more than 30 % gradient. However, extensive mechanisation imparts adverse impact on bush health and productivity as it leads to excessive banji shoot formation, dwarfing of crop with reduced intermodal length and leaf area affecting weight of crop shoots, magnesium, zinc and potassium deficiencies. To overcome the ill effects, UPASI recommends on foliar application nutrient mixtures for the benefit of South Indian tea plantations. It consists of N (2 kg), K (2 Kg), $ZnSO_4$ (2 Kg), $MnSO_4$ (500 ppm), $MgSO_4$ (2 kg), B (100 ppm), KNO_3 (4 Kg) NAA (10 ppm) and GA_3 (100 ppm) and amino acid mixture @ 300 ml all dissolved in 600 litres to spray one ha is known to increase the yield by 22-24 %.

Yield

Yield of made tea per hectare depends upon many factors such as elevation, clonal or seedling jats, management practices, severity of pruning, processing techniques etc., generally, in tea industry, a field which yields up to 2000 kg of made tea/ha is considered as low yielding and 2000 to 3000 kg as medium yielding and anything above 3000 kg as high yielding fields.

Organic tea refers to production of tea in the absence of synthetic chemicals and experience of many organic gardens is lower yield only and hence found not economical in view of higher labour costs. It may be viable provided organinc tea fetches 50-75 % higher price than normal one.

Importing countries insists on improvement in field practices, manufacturing processes, adopting high standard of hygiene and safety measures duly certified by agencies like Rain Forest Alliance, HACCP, etc which helps to meet the global competition from other counties.

Manufacturing of Tea

Basically, there are two types of processing viz., 1. **Orthodox method** in which the rolling operation is done in a series of rollers. The rollers have rotary tables with battens, jacket for loading the leaf and a pressure cup. 2) **CTC method** (cutting, tearing and curling) which has a CTC machine, consisting of series of a pair of rollers mounted in such a way they rotate in opposite directions and the clearance between them is so adjusted to crush and tear the leaves. Irrespective of the method, manufacturing of tea involves the follwing steps:

1. **Withering:** The objective of withering is to reduce the moisture content of leaves by spreading them in troughs which receive artificial air from fan fitted on one end. At the end of withering, the leaves attain a flaccid condition for which it may take 12 to 18 hours depending upon the weather condition.

2. **Rolling:** This operation is carried on by a series of machines or in a single roller, during which the cells in the leaves are broken to liberate the sap containing the polyphenol oxidase, an enzyme, which in the presence of oxygen, oxidises the polyphenols to produce **theaflavins** and **thearubigens.** These are responsible for colouring of the tea and are a prerequisite for next process viz., fermentation. Rolling takes place for 30-80 minutes. Afterwards, the fine sifted rolled ones are sent for fermentation while the coarse ones are again sent for rolling.

3. **Fermentation:** Rolled tea materials are either spread in concrete floors or kept in aluminium drums. In the presence of high humidity, air and proper temperature, the properly fermented tea will take golden red colour. This step decides the quality i.e. strength, colour and briskness of tea. Fermentation requires one hour or 2 hours depending upon the environmental conditions.

4. **Drying:** This step aims at stopping the fermentation process and slowly removing the moisture content without a burnt smell but preserving the inherent quality. This is achieved by passing the fermented tea in thin layers through conveyors into a drier in which the inlet temperature is maintained around 250-280 °F and outlet temperature is a round 150-200 °F. Proper drying takes 30-40 minutes.

5. **Grading:** Before grading, the dried tea is removed of the stalky fibres, which affect the quality, by passing through fibre separater machines. The bulk tea is passed through different sized meshes which aid in separation into different grades.

	Orthodox grades	**Mesh size**	**CTC Grades**	**Mesh size**
1.	Pekoe	>8 mesh sieve	Flowery Pekoe (FP)	> 8 mesh
2.	Tippy golden Orange pekoe (TGOP)	8-12	Pekoe	8-10
3.	Broken orange pekoe (BOP)	12-16	BOP	10-12
4.	BOP-Fannings	16-18	Pekoe-Fannings	12-16
5.	BOP-dust	18-24	BOP-Fannings	16-20
6.	Dust-I	25-30	Pekoe-Dust (PD)	20-30
7.	Dust- II	Below 30	Red-Dust (RD)	30-40
8.			Super Reddust (SRD)	40-50
9.			Fine Dust (PD)	50-60
10.			Superfine Dust (SD)	below 60

Plant Protection: Many pests and diseases are known to infect tea bushes and cause economic losses. The important pests and diseases, their typical symptoms and control measures are alone furnished below:-

	Pest	Symptoms	Control measures
1.	Red coffee borer (*Zeuzera coffeae*)	Larva borers the young stem, tunnels downward	Affected stems are cut up to the healthy and Indoxacarb 14.4 % (1.5 ml/L)is poured in the hole using ink filler and plugged with clay paste
2.	Phassus borer (*Sahydrassus malabaricus*)	Larva bores thick branches and makes long cylindrical tunnels, plugging the entry hole with chewed wood and silk	Affected stems are cut up to the healthy and Indoxacarb 14.4% (1.5 ml/L)is poured in the hole using ink filler and plugged with clay paste
3.	Shot-hole borer (*Xyleborus fornicatus*)	Grubs make a typical shot-hole on the branches and inside galleries. A serious problem in low and mid elevation areas	Deltamethrin 2.8 EC @ 500 ml/ha or Lambdacyhalothrin 5 EC @ 250 ml/ha or Emamectin benzoate 5 SG @ 0.009 % Two rounds of entomopathogens, *Beavaria bassiana* @ 1.5 kg/ha are recommended for spraying during end May and end Octoberwhen the humidity level is very high.
4.	Red spider mite (*Oligonychus coffeae*)	Infests upper surface of mature leaves	New acaricides such as Propargite 57 EC, Fenpyroximate 5 % EC, Fenpyroximate 5 % SC, and Hexythiazox 5.45 EC mitigate 5 EC are found to be more effective and their efficacy lasted for more than three weeks after application.
5.	Scarlet mite (*Brevipalpus californwus*)	Discolouration of leaves often leads to defoliation	
6.	Purple mite (*Calacarus carinatus*)	Leaves exhibit smoky grey colour	1. Propargite 57 EC @ 500 ml/ha 2. Fenpyroximate 5 EC/SC @ 300 ml/ha
7.	Pink mite (*Acaphylla theae*)	Young leaves turn pale and get twisted.	3. Hexythiazox 5.45 EC @ 400 ml/ha. 4. Paraffinic oil @ 1000 ml/ha 5. Lime sulphur - Polysulphide-S content 13-15 % @ 1:40, 10-12 % @ 1:30; 7- 9 % @ 1:25 and < 7 % @ 1:20 6. Neem Kernel Aqueous Extract 5 %(NKAE) could also be incorporated in to the schedule for RSM control.

			7. Abamectin 0.3 ml/L Paraffinic oil (horticultural spray oil) is currently recommended for pest control in various crops. Paraffinic oil has also been found effective against red spider mite, pink mite, purple mite, and thrips.
			Neem Kernel aqueous extract 5 % is found very effective in controlling RSM under field conditions and the efficacy is comparable to that of the synthetic acaricide.
			The entomopathogenic fungus (UPASI strains) *Paecilomyces fumosoroseus* has been found effective in controlling Red Spider Mite.
8.	Yellow mite (*Polyphagota-rsonemus latus*)	Infest pluckable shoot, leaves become rough, brittle and corky in under surface.	Propargite 57 EC @ 1 ml/L
9.	Thrips (*Scirtothrips bispinosus*)	Leaf surface becomes uneven, curly and metty, exhibiting parallel lines of feeding marks on either side of the midrib	Fibronil 5 SC @ 1 ml/L
10.	Nematodes (*Meloidogyne javanica, M. incognita*)	Occur in tea nursery, infested roots develop galls.	Pre heat treatement of soil media up to 60-80°C and application of carbofuran 3G @ 80g/cubic metre of medium.
11	Tea mosquito bug (*Helopeltis theivora*)	Sucks the saps from the tender leaves and causes crinkling of leaves, recently reported to cause 10-80 % loss in plantations	1. Fipronil 5 SC @500 ml/ha, 2. Thiamethoxam 25 WG @100 g/ha, 3. Deltamethrin 2.8 EC @ 500 ml/ha and 4. Bifenthrin 8 SC @ 500 ml/ha to contain the population of tea mosquito.

Diseases

1.	Blister blight (*Exobasidium vexans*)	Infects tender leaves and stem and develops translucent spot. Cloudy and wet weather favour infection.	Copper oxy chloride 350 g in 67 L of water with power sprayer for pruned field at 3-4 days interval. In the plucking field 210 g copper Oxychloride + 210 g Nickel chloride in 45 L of water/ha at 7 days interval. Recent recommendation is : **Fields under plucking:** Hexaconazole + Copper oxychloride 200ml + 210g (7 days) Propiconazole + copper oxychloride 125 ml + 210 g (7 days) **Fields recovering from pruning:** Hexaconazole + Copper oxychloride 200 ml + 210 g (5 days) Propiconazole + Copper oxychloride 125 ml + 210 g (5 days)
2.	Black root diseases (*Rosellinia arcuata*)	Infested roots show black mycelium on the roots, white star shaped mycelium between bark and wood and Black lead shot like perithecia seen on collar region.	The soil may be drenched with Dithane M45 @ 30 g/10 L
3.	Red root disease (*Poria hypolateritia*)	Infected roots exhibit blood red mycelium on washing. It spreads fast but slowly kills.	1. Take trenches of 1.2 m deep and 45 cm width surrounding the infected bushes and uproot and burn the bushes *in-situ*. 2. Rehabilitate soil with gautemala grass. The soil may be drenched with dithane M45 @ 30 g/10 L
4.	Brown root disease (*Fomes noxius*)	Infected root wood turns soft and spongy; it spreads slowly but kills quickly.	
5.	Root splitting disease (*Armillaria mellea*)	Infected roots develop crack, showing shoe-lace like rhizomorphs	

10

Coffee

COFFEE (*Coffea arabica* L. and *C. canephora* Linden)

Family: **Rubiaceae**

Coffee, native of Ethiopia, was introduced into India sometime during 1600 AD by a Muslim pilgrim, Baba Budan on the hills near Chikmangalur. Coffee seedlings were then planted in the backyards and it was not until the late 1820s that commercial plantations were started in South India by British entrepreneurs. Now coffee cultivation is mainly confined to the States of Karnataka, Kerala, Tamil Nadu and Andhra Pradesh and on a limited scale to Arunachala Pradesh, Assam, Madhya Pradesh, Manipur, Meghalaya, Mizoram, Nagaland, Orissa, Sikkim, Tirupura and West Bengal. The present area under coffee is 4,15,341 hectares of which arabica accounts for 49.50 percent and robusta 50.50 percent with a total production of about 3,18,200 tonnes. Production has grown 15 times since 1950 raising the national productivity to 852 kg/ha from 204 kg/ha. Among the total production, Arabica accounts for 32 % while robusta constitutes the balance (68 %). About 60 percent is exported annually earning around Rs. 300 crores and rest consumed internally. Italy is the major importer of Indian Coffee. There are 2,20,825 registered growers of whom 98.8 percent are small growers with 10 hectares and below. This plantation employs about 6,06,702 workers.

Botany: Though the genus *Coffea* consists of about 70 species, only two species are of economic importance. They are *C. arabica* (Arabica coffee) and *C. cenephora* (Robusta coffee) Arabica coffee, the fine-flavored, aromatic type makes up 60-65% of the total production and is also highly priced. The important difference among these two species is:

	Character	*C. arabica*	*C. canephora*
1.	Ploidy	Tetraploid ($2n = 44$)	Diploid ($2n = 22$)
2.	Adaptability	Higher elevation	Lower elevation
3.	Plant status	A small tree, a shrub or a bush under training.	A bigger tree than arabica.
4.	Leaves	Dark green	Pale green
5.	Blossoming after the receipt of blossom showers.	bloom in 9-10 days	bloom in 8-9 days
6.	Flowers	Scaly, small bracts, per axil 4-5 inflorescence of 1-4 flowers per inflorescence.	Leafy and expanded bracts with 5 to 6 flowers per inflorescence.
7.	Berries	10-20 per node oblong to round in shape.	40-60 or more per node, small.
8.	Fruit development period.	8-9 months	10-11 months
9.	Root system	Small but deep.	Large but shallow
10.	Pollination and Fertilisation	Self pollinated and Self-fertile	Cross pollinated and Self-sterile

The coffee plant has a prominent vertical stem with horizontal primary branches arising from it in pairs opposite to each other. Another upright shoot, sucker, arises from the main stem especially in a matured coffee plant in between the primary lateral branch and the leaf or its scar. It grows vertically like the main stem. These primary branches give rise to laterals (secondary) which in turn produce tertiary and quaternary branches. The secondary and tertiary types arise towards the distal end of the branch just above the axil (extra axillary) and the other type known as axillary bud which grows in the leaf axil and is capable of growing into a flower cluster or a lateral shoot. The axillary bud provides the main cropping wood for the plant (Fig. 15).

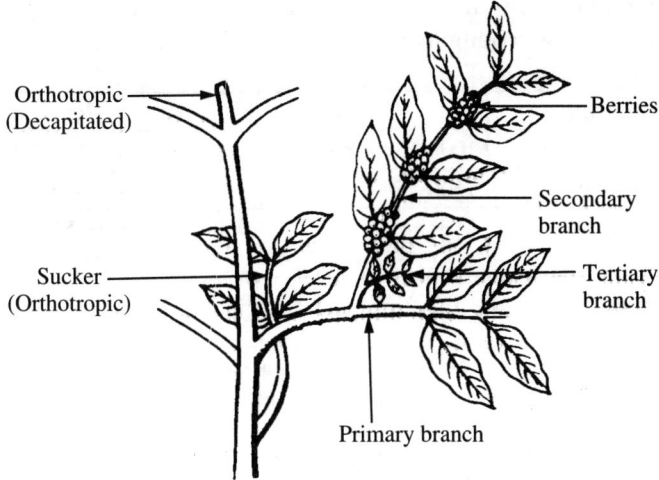

Figure 15   Morphology of a coffee plant

Coffee is a short day plant and in South India, flower initiation takes place between September to December. The flower buds grow into a definite size under fairly cold winter conditions and undergo a period of dormancy due to the onset of drought, coupled with high temperature, long day and high light intensity conditions prevailing in dry months i.e. November to March depending upon the places. There is practically no vegetative growth during these dry months. Immediately after the blossom showers, growth changes are conspicuous in flower buds on the third day following rains due to the moistening of the flower buds, soil wetness and low temperature that follows immediately after rain. This causes the plants to blossom within 7 to 10 days. This imposed dormancy is a necessary event, as it enables single harvest, otherwise, coffee will be blooming throughout the year resulting in staggered harvesting concomitant with increased cost of picking, etc. The fruit is a drupe and normally contains two seeds. Abortion of one ovule due to non- fertilization leads to the formation of a single seeded fruit, called pea berry. Sometimes, 3 or more seeds may be present, due to trilocular ovaries or false polyembryony and is often called triangular seeds. Occasionally, formation of more than one ovule per locule is seen and such seeds are known as elephant bean.

Climate and Soil: Climatological factors like rainfall, temperature, elevation and aspect can influence economic production of coffee much more than soil factors. Soil should be deep, well drained, slightly acidic in reaction and rich in organic matter content.

The optimum soil and climatic requirements for arabica and robusta under South Indian conditions are as follows:

	Particulars	Arabica	Robusta
1.	Elevation	1000-1500 m MSL	500-1000 m MSL
2.	Annual rainfall	1600-2500 mm	1000-2000 mm
3.	Blossom rain	March-April (2.5 to 4.0 cm)	February-March (2.0 to 4.0 cm)
4.	Backing rain	April-May (5 to 7.5 cm)	April-May (5.0 to 7.5 cm)
5.	Shade	Needs medium to light shade depending on elevations and aspects.	Needs uniform thin shade.
6.	Temperature	15-25°C (ideal) (cool equable)	20-30°C (ideal)
7.	Relative humidity	70-80 %	80-90 %
8.	Soil	Deep friable, porous, rich in organic matter moisture retentive, slightly acidic pH 6.0 to 6.5	Same as for Arabica
9,	Aspect	Northern, Eastern and N.Eastern aspects are ideal.	Flat to gentle slopes
10.	Slopes	A gentle to moderate slope is ideal	Gentle slopes to fairly level fields are to be preferred.

Coffee cultivation is confined mostly to the hilly tracts of Western and Eastern Ghats. A well distributed rainfall is preferable for coffee with dry months from December - March. Summer showers are important for flowering and failure of blossoms showers leads to crop loss.

Varieties: Crop improvement work carried out at Central Coffee Research Institute, Balahanur, Karnataka has resulted in the release of a number of superior selections in Arabica coffee and the popular ones are furnished below:

	Name of the variety	Parentage	Special attributes
1.	S.795 (Sln. 3)	A cross between S.288 × Kent	Resistant to leaf rust race 1 and 11, popular among growers. Yield up to 2000 kg/ha

2.	Sln 5B	Devamachy × S.333	Exhibit higher resistance to leaf rust. Produce bold fruits and a Yield potential of 1500 kg/ha
3	Sln. 7 (San Ramon hybrids)	San Ramon-a short internode arabica type-spotted in Costa Rica.	Dwarf in stature, but segregates to tall by 30 % also. Drought resistance with a yield potential of 1500 kg/ha
4	Sln. 8 (Hibrido-de-Timor) commonly called as HDT	A spontaneous hybrid of robusta-arabica origin, spotted in Portugese Timor island.	Highest vertical resistance to leaf rust, phenotype and bean quality resembles arabica. Moderate yielder –1200 kg/ha
5	Sln. 9	HDT (Sln. 8) × Tafarikela	Drought hardy, suitable to different coffee zones. More resistance to leaf rust with yield potential up to 1500 kg/ha
6	Sln. 10 (Catura crosses other than HDT)	Caturra crossed with Cioccie and S.795. The respective progenies were again crossed to evolve a double cross hybrid.	High yielding potential but highly susceptible to rust.
7.	Cauvery	F4 cross between Caturra × HDT	Comes to bearing within three years, yield potential 2.5 t/ha, amenable for close planting.
8.	Chandragiri	Villa Sarchi × HDT	Semi dwarf variety suitable for all agro climatic zones. Produces relatively bolder beans compared to other arabicas. Yield potential around 1750 kg/ha

Nursery: Healthy and mature fruits of normal size and appearance, three quarters to fully ripe are harvested from specially selected and marked coffee plants for use as seed bearers. Floats are discarded, the sound fruits are pulped, the beans drained and sieved to remove defective beans. The beans are then mixed with sieved wood-ash, evenly spread out to a thickness of about 5 cm and allowed to dry to facilitate uniform drying. Excess ash is rubbed-off after five days of drying.

Germination beds, raised to a height of about 15cm, one metre width and of convenient length are prepared. Four baskets of fully mature cattle manure or compost, about 2 kg. of finely sieved agricultural lime and 400 g of rock phosphate are incorporated in a bed measuring 1 m × 6 m.

Seeds should be sown with the flat side facing the soil at a distance of 1.5 - 2.5 cm from one another in regular rows. A thin layer of fine soil is then spread. The bed is covered with a layer of about 5 cm of paddy straw. The beds are watered daily and protected from direct sunlight by an overhead pandal constructed for this purpose. The seeds germinate in about 45 days. The seedlings are then transplanted to secondary nursery beds or raising polybag plants.

Coffee seedlings are transplanted to polythene bags of 23 cm × 5 cm with 150 gauges thick in February or March when they are at the button or topee stage. The bags are filled with a prepared mixture of 6 parts of jungle soil, 2 parts of well rotten sieved cattle manure and 1 part of fine sand. At the time of transplanting it is preferable to slightly nip the tap root of the seedling. Transplanting is done preferably in the early morning hours or late in the afternoon.

Regular watering and after-care of the seedlings should follow. Seedlings may be manured once in 2 months with urea dissolved in water, 20 g urea in 4.5 litres of water is sufficient for an area of 1 square metre. Adequate protection is given against nursery diseases and pests. Overhead shade in the nursery has to be thinned and finally removed after the onset of monsoon.

Preparation of land: Clean felling is not advocated. Selective retention of desired species of wild shade trees is essential. The land should be divided into blocks of convenient size with foot path and roads laid out in between. In steapy area, terracing and contour planting may also be adopted.

Spacing for arabica and robusta coffee is 2 to 2.5 m and 2.5 to 4 m respectively on either way. A close planting at 1-1.5 m either way and reduce the population by half after one or two harvests is good. Pits of 45 cm are usually opened after the first few summer shower and seedlings of 16 to 18 months old are planted during June or September - October. A hole is made in the centre of the pit after levelling the soil. The seedling is placed in the hole with its tap root and lateral roots spread out in proper position. The hole is then filled. The soil around the seedling is packed firmly and evenly in such a way that 3 cm high above the ground to prevent stagnation of water around the collar. The seedlings are provided with cross stakes to prevent wind damage.

Training and Pruning: Training of the bush is necessary to have a strong frame work which promotes production of bearing wood.

Coffee is trained in two systems viz

1. **Single stem system:** When the plant reaches a height of 75 cm in Arabica or 1.10 to 120 cm in robusta, lit is topped. This helps to restrict vertical growth, facilitate lateral spreading and increase the bearing area. In this system, a second tier is also allowed sometimes depending upon the soil fertility and plant's vigour.

2. **Multiple stem system:** which is common in Kenya, Tanzania, is not practiced in India.

Pruning in coffee is generally done immediately after harvest and till the onset of monsoon. It is essentially a thinning process and is done mainly to divert the vigour of the plants to certain parts by pruning the other parts. Pruning involves (a) centering-removal of the vegetative growth up to 15 cm radius from the centre and upto the first node of all primary branches (b) de-suckering - removal of orthotropic branches arising from the main trunk (c) handling - removal of small sprouts arising from the axis of the leaves which otherwise grow towards the inner side and cause shade and become unproductive wood and (d) nipping - growing tip of primary branches is removed to encourage secondaries and tertiaries.

Soil Management

Soil management practices aim at conserving soil and water and in general to make the soil perform its functions satisfactorily. It includes the following practices in coffee.

(a) **Digging:** In the new clearing, the field is given a thorough digging to a depth of about 35 to 45 cm towards the end of the monsoon. All weeds and vegetative debris are completely turned under and burned in the soil while the stumps are removed. Once the coffee plants have closed in, annual digging is not done.

(b) **Scuffling or soil stirring:** In established coffee fields, scuffling or soil stirring is done towards the beginning of the dry period. It controls weeds and also conserves soil moisture.

(c) **Trenching:** Trenches and pits are dug or renovated in a staggered manner between rows of coffee along the contour during August-October when the soil is fairly easy to work. These are 50 cm wide and 25cm deep and can be of any convenient length.

(d) **Mulching:** Mulching young coffee clearings helps to maintain optimum soil temperature and conserve soil moisture and acts as an effective erosion control measure. Mulching also adds to fertility of the soil.

(e) **Weed control:** New clearings are hand-weeded three to four times a year and established coffee two to three times. During the monsoon, the weeds are slashed back. Another weeding is done towards the end of the monsoon. Clean weeding is generally done during the post-monsoon period.

   Of late, chemical weedicides have gained popularity in larger plantations. Gramoxone at 1.25 litres in 450 litres of water per hectare has been found to be the best. This should follow in slash weeded plots after 10-15 days. Glyphosate 41 EC (Round up or Glycel) 3 litres in 450 litres of water can also be used.

(f) **Irrigation:** Sprinkler irrigation is mainly used as an insurance against failure of good blossom or backing showers. It is also used on young plantations, marginal areas where water is available in plenty to help in establishment of coffee and shade.

(g) **Soil acidity and liming:** The heavy rainfall in coffee growing zones of South India brings about leaching in calcium and magnesium leading to soil acidity. Besides, continuous use of acid forming fertilizers like ammonium sulphate also makes the soil acidic. As the ill effects of soil acidity are more, periodical lime application is essential to correct the soil pH for good productivity. The quantity of lime to be applied is based on soil pH and lime requirement.

Agricultural lime and dolomitic lime are the most commonly used liming materials. It can be applied to the soil at any time during the year provided there is a gap of one month or a few showers between lime and fertilizer applications. It is also desirable to apply lime when there is sufficient moisture in the soil for quick response.

Liming may be carried out during May-June and November-February in the South-West monsoon areas and January to March and June-July in North-East monsoon areas. It is broadcast in between rows of coffee and incorporated into the soil by light digging or forking.

Shade and its Management

Under the climatic conditions existing in India, Coffee is being cultivated under shade. It comprises of two canopies-lower or temporary and upper or permanent. High light intensities and temperature prevailing during the drought period are not conducive for normal and healthy growth of coffee plants. Therefore, there is necessity for protecting the coffee plants during the above period by providing both temporary (lower) and permanent (upper) shade trees.

Dadap *(Erythrina lithosperma)* is used as a lower canopy shade in India. It is planted along with coffee in new clearings. When stakes are planted in June, they grow quickly using the moisture available in the soil. Next to dadap, silver oak *(Grevillea robusta)* is the most commonly used tree for temporary shade. It grows easily and rapidly on new clearings but does not give much leaf mulch. The most popular permanent shade trees found in South India are *Albizzia lebbek, A. moluccana, A. odoratissima, Artocarpus integrifolia, Cedrella toona, Dalbergia latifolia, Ficus glomerata, F. infectoria, F. retusa and Maesopsis eminii.*

Permanent shade trees are generally planted about 12 to 14 metres apart. It is advisable to plant a large number initially and thin out as the trees grow and spread out. The trees have to be regulated in such a way that in course of time, they have their canopy about 10 to 14 metres above the coffee. Shade trees require constant attention by way of pruning and lopping to provide the required filtered shade to coffee. The most convenient time to regulate shade is after pruning and liming.

The dadaps are lopped 2 to 3 times a year to regulate the light requirements according to season. Dadap is lopped at the commencement of the South West monsoon to allow more light and regulated during winter when the dry weather begins. In North East monsoon area, it is lopped during August-September.

Manuring

Coffee plants produce every year fresh wood for the succeeding crop concomitant with the function of maturing the current berries. Hence, they require a regular supply of nutrients. Besides, being grown in heavy rainfall area, the losses of nutrients due to leaching and fixation are to be offset by regular application of adequate quantities of fertilizers.

The peak periods of demands for nutrients are at the time of flowering, fruitset and development and maturation of the crop. Based on all these factors, Coffee Board recommends the following dose of fertilizers for coffee (Table 2).

The leaf mulch beneath the coffee is swept towards the base and the fertilizers are applied in a broad circular band about 30 cm away from the stem. They are then incorporated into the soil, with a fork or stick and covered by mulch.

Table 2 Manurial recommendation for Coffee

	Pre-blossom March N:P₂O₅:K₂O	Post blossom May N:P₂O₅:K₂O	Mid-monsoon August N:P₂O₅:K₂O	Post-monsoon October N:P₂O₅:K₂O	Total
ARABICA					
Young coffee 1st year after planting	15:10:15	15:10:15	...	15:10:15	45:30:45
2nd and 3rd year	20:15:20	20:15:20	...	20:15:20	60:45:60
4th year	30:20:30	20:20:20	...	30:20:30	80:60:80
Bearing coffee 5 years and above for less than one tonne/ha. crop	40:30:40	40:30:40	20:0:0	40:30:40	140:90:120
for one tonne/ha. and above	40:30:40	40:30:40	40:30:40	40:30:40	160:120:160
ROBUSTA					
for less than 1 tonne/ha. crop	40:30:40	...	...	40:30:40	80:60:80
for 1 tonne/ha. and above	40:30:40	40:30:40	...	40:30:40	120:90:120

As a supplement to soil applications of fertilizers, foliar spraying with (Urea 0.5 kg, ammophos (20:20) 0.5 kg and muriate of potash 350 g dissolved in 200 litres of water or Bordeaux mixture may be given) during periods of slow growth, flowering and fruit-setting. However, Bordeaux mixture should be neutralised properly before dissolving the nutrients.

Harvesting

Coffee fruits should be picked as and when they become ripe to get better quality. Arabica comes for harvesting earlier since they take 8-9 months for fruit development from flowering while robusta takes 10-11 months. Picking is done by hand. The first picking consists of selective picking of ripe berries often seen in the outer portion of the node and is called fly picking. Thereafter, there will be 4-6 main pickings at 10-15 days intervals and final harvest i.e., stripping consists of picking of still remaining green berries on the plant.

Processing: Coffee is processed in two ways a) Wet processing to prepare plantation or parchment coffee and b) dry method by which cherry coffee is prapared.

A. Preparation of Parchment Coffee

(i) **Pulping:** This method requires pulping equipment and adequate supply of clean water. Fruits should be pulped on the same day to avoid fermentation before pulping. Fruits may be fed to the pulper through siphon arrangements to ensure uniform feeding and to separate lights and floats from sound fruits. The pulped parchment should be sieved to eliminate any unpulped fruits and fruit skin.

The skins separated by pulping should be let away from the vats into collection pits so that microbial decomposition of the skin will not affect the bean quality when it gets mixed up with the bean.

(ii) **Demucilaging and washing:** The mucilage on the parchment skin can be removed by

(a) **Natural fermentation:** The mucilage breaks down in the process of fermentation and it takes 24 to 36 hours for arabica and 72 hours for robusta. Cool weather delays the process of fermentation. Under fermented or over fermented beans affect quality. When correctly fermented the mucilage comes off easily and the parchment does not stick to the hand after washing and the beans feel rough and gritty when squeezed by hand. When the mucilage breakdown is complete, clean water is let in and the parchment washed pebble clean with three to four changes of water.

(b) **Treatment with alkali:** Removal of mucilage by treatment with alkali takes about one hour for arabica and one and a half to two hours for robusta. The beans obtained after pulping are drained off excess water and spread out in the vats uniformly and furrowed with wooden ladles with a long handles. A 10 % solution of caustic soda (sodium hydroxide) is evenly applied into the furrows using a water can. 10 litres of the alkali is sufficient to treat 25 to 30 forlits (One forlit = 40 litres) of parchment. The parchment is agitated thoroughly by the ladles so as to make the alkali to come into contact with the parchment and trampled by feet for about half an hour. When the parchment is no longer slimy and makes a rattling noise, clean water is let in and the parchment washed clean with three or four changes of water.

(c) **Removal of mucilage by friction:** There are machines which pulp and demucilage the beans in one operation. However, a number of naked and bruised beans may result in the parchment. It is, therefore, necessary to adjust the machines carefully to obtain uniform pulping and demucilaging. Cup-test results have indicated that there is no difference in cup quality coffee processed by different method.

(iii) **Drying:** The next stage is drying the parchment in the sun until the moisture content is sufficiently reduced to permit storage of beans till they are despatched to curing works. Proper drying contributes to the healthy colour of the bean and other quality factors. Under-dried parchment turns mouldy and gets bleached during storage and subsequent curing operations.

The parchment is spread on clean tiled or concrete drying floor to be dried slowly by spreading to a thickness of about 7 to 10 cm. Stirring and turning over coffee, at least once an hour, is necessary to facilitate uniform drying. The parchment should be heaped up and covered in the evening until next morning. Sun drying may take about 7 to 10 days under bright weather conditions. At the right stage of dryness the parchment becomes crumbly and the beans split clean without a white fracture when bitten between the teeth. Drying is complete when a sample forlit of coffee records the same weight for two days consecutively. At this stage, coffee is shifted to the stores and bagged in clean, new gunnies.

When coffee is being dried, all naked beans, pulper nipped and bruised beans, blacks, greens and other defective beans are sorted out and despatched to curing works separately.

B. Preparation of Cherry

For preparation of cherry coffee, fruits should be picked as and when they ripe. Greens and under-ripe fruits should be sorted out and dried separately. The fruits should be spread evenly to a thickness of about 8 cm on clean drying ground in which the cherries are stirred and ridged atleast once every hour. The cherry is dry when a fistful of the drying cherry produces a rattling sound when shaken and a sample forlit records the same weight on two consecutive days. The cherry should be fully dry at the end of 12 to 15 days under bright weather conditions.

Plant Protection: The important pest and diseases in coffee, their symptoms/damage and the control measures are presented below.

	Pests/diseases	Symptoms	Control measures
Pests			
1.	Mealy bug (*Planococcus lilacinus* and *P.citri*)	Serious foliar para sites, attack starts from a few isolated bushes and then spreads to others	Spray the affected patches with 4 litres of kerosene in 200 litres of water along with 200 ml of an agricultural wetting agent. Care should be taken to thoroughly emulsify kerosene with the wetting agent before adding to the barrel and stir the solution frequently to avoid separation of the oil. If the root is affected, drench the root zone with dimethoate 30 EC at 0.1 % a. i. (about 660 ml in 200 litres of water) Release the parasitoid *Leptomastix dactylopii* against *P. citri* or the predator, *Cryptolaemus montrouzieri* irrespective of the species of mealy bugs.
2.	Green bug (*Coccus viridis*)	Serious foliar para sites, attack starts from a few isolated bushes and then spreads to others	Spray affected patches with dimethoate 30 EC (Rogor) at the dosage of 170 ml in 200 litres of water with 200 ml of any wetting agent.

| 3. | White stem borer (*Xylotrechus quadripes*) | Plants show unhealthy signs like wilting and yellowing of leaves | Provide good shade, burn the infested plants in situ, bark scrubbing to remove the loose barks and smoothening the surface of the stem to discourage egg laying, Spraying 10 % lime (spray lime at 20 kg in 200 litres of water along with 200 ml Fevicol DDL) on main stem and thick primaries and Spray the main stem and the thick primaries once during mid April and once during end October or first week of November. Since the winter flight period is of long duration, one more spray may be given in the first week of December with Chlorpyriphos 20 EC @ 600 ml in 200 litres of water along with 200 ml of any wetting agent. Pheromone trap can be installed on coffee estates for monitoring and trapping the beetles during the peak flight periods (April-May and October-December) to reduce the infestation. |
| 4 | Coffee berry borer (*Hypothenemus hampei*) | Damage is caused to young as well as developed berries. Breeding occurs in developed berries from the time the endosperm becomes hard. It may continue in the black left over ripe berries either on the tree and or on the ground. The infested berry can be easily identified by the presence of small holes, | The population of this pest could be effectively maintained at very low levels by adopting an integrated pest management programme which includes cultural, biological and chemical measures. Timely and thorough harvest is also important to reduce the pest build up. At the time of harvest, spreading polythene sheets or mats under the plants is very useful in minimizing the gleanings and improving the harvesting efficiency. Infested fruits should be processed after dipping in boiling water for 1-2 minutes. Heavily infested berries may be destroyed by burning or burying in the soil to a depth of at least 20 inches. Maintain thin shade |

		generally one and sometimes two or three, in the navel region. The berry borer attacks all the coffee cultivars.	and train the plant regularly so that harvest and plant protection operations are efficient. Chlorpyriphos 20 EC at the dosage of 600 ml per barrel along with 200 ml of any wetting agent is also very effective and can be used for hot spot spray. The white muscardine fungus *Beauveria bassiana* could be used as a bio control agent against this pest very effectively.
5.	Shot-hole borer *(Xylosandrus compactus)*	Attacked plants dry up, extensive tunneling within the branches seen	Prune and burn the affected branches
6.	Cockchafers *(Holotrichia spp.)*	Grubs feed on feeder roots of coffee, old plants withstand but young ones often wilt and die.	In white grub-infested areas, incorporate 5 g of Thimet 10 G or 10 g of Furadan 3 G into the pits at the time of planting. Collect and kill the grubs encountered while taking up digging and other farm operations. Install light traps after the first summer showers during March-June and kill the trapped adults.
7.	Nematodes *(Pratylenches coffeae)*	General stunting, yellowing of leaves, distortion of roots leading to die-back, affected plants easily get dislodged due to poor anchorage	**In the nursery** Dig up the nursery site and expose the soil to the sun during summer. Sieve and dry jungle soil and farmyard manure thoroughly before use. Avoid obtaining nursery plants from infested areas. **In the field** Uproot and burn the affected plants. Dig up pits and expose the soil to the sun for one year. Plant the area with robusta (if suitable) or arabica –robusta grafted plants (arabica scion grafted on to robusta root stock). Drench the soil around the plants with carbosulfan 25 EC (Marshal) @ 480 ml in 200 litres

			of water. Depending on the size of the plant, 50 to 500 ml of the solution would be required per plant.
Diseases			
1.	Leaf rust *(Hemelia vastatrix)*	Pale yellow spots on the lower surface of leaves, later turn to orange yellow powdery mass, infected plant exhibit defoliation	Spray Bordeaux mixture (0.5%) pre blossom (Feb - Mar), Pre monsoon (May-June), and post monsoon months (Sep - Oct) Bayleton (160 g in 200 lit of water) or Contaf (400 ml in 200 lit of water) Premonsoon and Post monsoon.
2.	Blackrot *(Koleroga noxia)*	Blackening and rotting of the affected leaves, twigs and developing berries	Proper shade regulation, centering and handling the affected bushes to prevent secondary spread, spraying with 1 % Bordeaux mixture or spraying the endemic blocks with Bavistin (120 g in 200 lit of water) just before the onset of monsoon and during the break in monsoon.
3.	Brown blight, twig blight, dieback *(Colltetotrichum gloeosporioides)*	Small water soaked lesions on margins of leaves and slowly extend causing drying of margins, often bushes start drying downwards with shedding of berries	Prune badly affected plants during dry months, spray 0.5% Bordeaux mixture
4.	Root diseases *(Fomes noxius, Poria hypolateritia, Rosellinia bunodes)*	Affected plants show gradual yellowing of leaves, defoliation followed by death of above ground parts	Uproot the affected plant and burn, dig trenches of 60 cm deep and 30 cm width to isolate the affected bushes, keep fallow for 6 months, apply organic manures 10-15 kg per pit.

11

Rubber

RUBBER (*Hevea brasiliensis* Muell-Arg.)

Family: **Euphorbiacace.**

Natural rubber for commercial production is available from *Manihot glaziovii* (cera rubber), *Ficus elastica* (Indian rubber), *Castiolla elastica* (Panama rubber), *Parthenium argentatum* (Guayul), *Taraxacum koksaghyz* and *Hevea brasiliensis* (Para rubber) and among them; *Hevea brasiliensis* is the most important commercial source of natural rubber. It is native of Brazil and was introduced in Asia in 1876.

Indonesia, Thailand, Malaysia, China and India are the main rubber producing countries. In India it is cultivated in an area of 7,11,560 hectares with an annual production of 8,61,950 tonnes, ranking fifth in production. Around 109,77,000 MT of natural rubber is produced currently in the world and almost all of them are consumed by various countries. Similarly, nearly 150, 90,000 MT of synthetic rubber are produced globally to meet the demand of the rubber industry. In India although it is grown in many states, Kerala has nearly 75.08 % share by area (5, 43,228 ha) and 89.45 % share by production (77,05,80,000 Tonnes)

Botany

Rubber tree belongs to the natural order Euphorbiaceae. This tree is sturdy, tall and quick growing. It has a well developed tap root and laterals. The leaves are trifoliate, with long petioles. The young plants show characteristic growth pattern of alternating period of rapid elongation and consolidated development. Normal annual leaf fall known as 'wintering' occurs in the case of mature trees during December to February in South India. Flowering and foliation follow wintering. Off-season flowering also occurs during September-October. The flowers appear on the new leafy twigs as panicles. They are unisexual, small and fragrant. Staminate flowers are small and numerous. Pollination is by insects.The fruits mature in about five months after pollination. They are three seeded and burst when mature, scattering the seeds 15 to 18 meters away. The seeds weigh 4 to 6 g, they possess a hard brown coat having characteristic mottling. Seeds of seedling trees and different clones vary in size, shape, weight and seed coat markings. The seeds belonging to a clone have characteristic size, shape, weight and seed coat markings which are helpful in their identification.

Latex vessels are present in all parts of the tree except in the wood. These vessels originate from cells produced by cambium and they are arranged in rows. The vessels do not have straight vertical courses within the bark. Generally, they run in anti- clockwise direction, the angle of inclination varying from 3 to 5°. The latex vessels within the same ring are tangentially interconnected forming a mesh-like structure. The latex vessels are thin walled and towards the hard bark region they become shriveled, discontinuous and non-productive. Latex vessels in the stem and branches show the same structure. The number of latex vessels, however, is comparatively less in the branches.

Climate and Soil

Rubber is exacting in its climatic requirements. The regions lying within 10° latitude on either side of the Equator is highly suitable for rubber cultivation. It requires a temperature ranging from 20° to 30°C with a well distributed rainfall of 200-250 cm over the year. It comes up in plains and also in slopes of mountainous regions ranging from 300-800 m above sea level. This specific climate is available only in Kanyakumari district, Tamil Nadu and Kerala which constitute the traditional area.

It thrives well in deep well drained acidic soils of red lateritic loams or clayey loams with a pH varying from 4.5 to 6.0.

Varieties

Clonal varieties have been developed by Rubber Research Institute of Malaya, Rubber Research Institute of India, Kottayam and other institutes. Clones are broadly classified into three categories viz., primary, secondary and tertiary, based on the method adopted for the development of their mother trees. When mother trees are selected from existing seedling populations of unknown parentage and are multiplied vegetatively to give rise to the clones, they are called primary clones. When the mother trees are evolved by cross pollination (hand pollination) between two primary clones and are then multiplied vegetatively, they are known as secondary clones. Tertiary clones are also produced by controlled pollination of two existing clones, but they differ from the secondary clones in that atleast one of parents or both the parents are of secondary clones.

Rubber Board of India recommends some of the following clones for cultivation in South India.

	Name	Parents	Important traits
Primary clones			
1.	TJIR-1	–	Indonesian clone, yield 930 kg/ha per year, 'S' to Phytophthora, Oidium and pink disease.
2.	G.T.1	–	Indonesian clone, yield 1360 kg/ha year, 'T' to Phytophthora, pink disease and brown bast.
3.	G.I.1	–	Malaysian clone, yield 1130 kg/ha year, 'S' to brown bast, possess drought tolerance.
4.	P.B.86	–	Malaysian clone, yield 1130 kg/ha per year.
Secondary clones			
5.	PRIM-600	TJIR – 1 × P.B.86	Developed by Rubber Research Institute of Malaya (RRIM). Yield 1317 kg/ha per year, 'S' to Phytophthora and pink disease.
6.	RRIM. 628	TJIR.1 × RRIM.527	Yield 1051 kg/ha per year, 'S' to brown bast, poor yielder during summer.
Tertiary clones			
7.	RRIM-703	RRIM; 600 × RRIM. 500	Yield 1725 kg/ha per year, 'S' to brown bast and wind damage.

(S = Susceptible, T = Tolerant)

Other important clones recommended for South Indian conditions are **PR.** 107, **PB.** 5/51, **RRIM**-118, **RRII**-203 and **RRII**-208.

It may be observed that all the desirable characters viz., high latex content may not be available in a single clone. Therefore, to get the benefits of mixed clones in a population, polyclonal seed gardens are being raised. Superior clones numbering three to six are planted in an isolated area and allowed for natural open pollination. The seeds so collected from such garden may produce good seedling families. Some of the common clones suitable for inclusion in polyclonal seed gardens are **RRIM**-600, **RRIM**-628, **GT.**l, **PB** 5/51, **PB** 28/59, **TJIR**.1 and other modern clones which produce dependable seedling families. Polyclonal seed gardens have been established in Kanyakumari and south Canara districts.The number of seeds harvested from a garden varies widely depending on various factors. However, an average of 150 seeds per tree can be expected from a properly maintained garden.

Propagation

Rubber is propagated by seeds and by budding.

1. **Seeds:** Propagation through seed is practiced to raise seedlings for rootstock purpose or to raise polyclonal seedling progenies. Seeds normally ripen during July-September in South India. As the viability is very short (8 weeks), they are to be sown immediately. Raised beds of river sand of 1 m width and of convenient length are formed and the seeds are sown in a single layer touching one another and pressed firmly with the surface of the seed just visible above. Nursery may be protected from direct sun by providing a temporary shade. Regular watering is attended to maintain the moisture in the beds. Seeds start germinating within 6 to 10 days. Such sprouted seeds may be picked and planted in the nursery at 30 × 30 cm to raise seedling stumps or at 60 × 90 cm or 60 × 120 cm to raise bud wood nursery or stumped budding. Otherwise, sprouted seeds can be directly planted in the field.

2. **Budding:** The scions of a particular clone are maintained in the bud wood nursery by planting the budded stumps or by budding the clone on the seedlings *in situ* at nursery. Budded stump often refers to the budded plant whose scion shoot is cut very close to the budding zone leaving few dormant buds in the scion shoot. On the other hand, if the root stock is cut as a stump and budding is done, usually green budding at four to five months stage, then it is known as stumped budding.

When the budwood nursery plants are one year old, about 1 m of usable budwood can be obtained. The budwood is cut when atleast 1m of brown bark has developed. The immature green portion should be removed to a point about 1 m below the terminal bud, leaving the leaf stalks in position. The budwood may be cut off about 15 cm at the base, leaving a few dormant buds to develop into bud shoots for the subsequent season. Two such sprouting shoots may be allowed for next year, from one metre shoot, 15 to 20 buds may be obtained.

Depending on the colour and age of buds, two types of budding are recognized-'brown budding' and 'green budding'.

Brown budding: It is carried out by grafting buds taken from budwood of one year growth on to stock plants of 10 months or more growth. Vigorous healthy stock seedlings having a girth of about 7.5 cm at the base are ideal for budding. In South India, budding is best done during the wet months. Budding can be done at any time of the stock seedlings and budwood peels easily.

Green budding: Both the stock plant and bud wood used for green budding are young. Vigorous seedling of two to eight months age, with a girth of about 2.5 cm and brown bark up to a height of about 15 cm from the base are used as stocks. Green buds are taken from bud shoots of 6 to 8 weeks growth harvested from specially raised source bushes or existing budwood nursery.

Modified forket method is followed and is done during April-May, when the weather is not dry or wet. Two types of budding techniques are practiced. Brown budding is done by using buds taken from bud wood of one year growth on to a stock plant of ten months old. Green budding on the other hand involves young green budwood and stock. Bud wood of 6-8 weeks old is used on stock seedlings of 2 to 6 months old.

Recently, polybag plants are raised as such plants reach tapping stage quickly. Black polythene bags of 60 × 30 cm with 400 gauges are filled with top soil alone along with 25 g of rock phosphate. Green budded stumps are planted in these polybags and the scions are allowed to develop 2 to 3 whorl of leaves. Further, root trainer plants are also used for planting. Root trainers have a length of 26 cm with a holding capacity of 600 cc. The specific features like tapering shape, vertical ridges in the container wall, drainage hole at the bottom are all incorporated with the purpose of properly training the structural development of roots and hence the container is termed as root trainer. Cured coir pith mixed with appropriate quantities of single phosphate, neem cake, bone meal, fungicides and pesticides is used as the potting medium. After planting budded stumps, preferably green budded stumps, the root trainers are stacked in carriers made of iron rod or bamboo splints. Initially the base of the containers is covered with top soil and the roots growing out of the drainage hole are permitted to grow into this soil. On attaining sufficient growth the soil put beneath the containers is removed, roots are pruned and the plants along with the containers are kept suspended in air for hardening for a minimum of eight weeks. During the hardening process the tap root resumes growth in a few days and undergoes natural air pruning near the hole at the bottom and thus prevent its coiling inside the container. This stress induces emergence of large number of lateral roots, in its turn, also undergo air pruning and as a result a hardened root trainer plant will have a root system consisting of a central tap root and large number of lateral roots well oriented without any deformity. The root plug is separated from the container just before transplanting to the field. The rot trainer plant is kept up right down and the brim of the containers is tapped against a hard surface so that the root plug comes out of the container without causing any damage to the roots. A planting hole is made in the refilled pit by pressing the empty container itself and the toot plug are inserted in to it. The soil is pressed slightly from the sides and all other cultural practices are adopted as in the case of polybags plants.

Planting

For new plantings, jungle clearing with felling of trees has to be done first. Pits are usually dug to the size of the one cubic metre and are filled up with soil and compost. Planting may be done in a rectangular or square or quincunx system during June-July.

The common spacing adopted for budded plants are as below:

In hilly areas	6.7 m × 3.4 m	445 plants/ha
In flat areas-Square system	4.9 m × 4.9 m	420 plants/ha
Triangular System	4.9 m × 4.9 m	470 plants/ha

The size of the pits varies with the nature of the soil. In soils with a depth of 1 m or more, planting can be done in small pits dug to accommodate polybag plants. However in hard soils planting should be done in larger pits of size 75 cm^3 or 90 cm^3 depending on soil hardness. Mechanized pitting using tractor-mounted hoe digging machines and pitting and terracing by earth movers are increasingly practiced.

After care

1. **Cover cropping:** Growing cover crops is important in rubber plantations to prevent soil erosion, conserve soil moisture, keep down the soil temperature and add mulch and organic matter to the soil. Some of the cover crops commonly used in South India are *Pueraria phaseoloides, Calopogonium mucunoides, Centrosema pubescens* and *Mimosa invisa* var. *inermis.*

   It is often desirable to establish a mixed cover to simultaneously get the benefits of these cover crops. A mixture of *Calopogonium, Pueraria* and *Centrosema* seeds in the ratio of 5:1:4 is used for sowing. In this mixed cover, *Calopogonium* grows much rapidly and covers the ground quickly during the *first* year itself. Then *Pueraria* and *Centrosema* start dominating. Dense and vigorous growth of *Pueraria* suppresses weeds, but it starts fading out when the canopy closes. Subsequently, *Centrosema* continues to grow under shaded conditions while the former two cover crops fail to thrive. Thus, complete benefits of cover crops starting from the first year of planting could be obtained if mixed cover crop is established in rubber plantations.

2. **Weeds and weeding:** The weeds can be eradicated either by labour or by employing weedicides. When weedicides are employed, care should be taken that the cover crops are not affected. 2, 4-D formulations (Fernoxone @ 2 kg in 450 litres of water) may be sprayed early in the season to eradicate the weeds. Post emergence herbicide paraquat @ 2.251 kg/ha? and 2,4-D Fernoxone @ 1.25 kg/ha can be mixed to control broad leaves and grass weeds if diluted in 500-600 litres of water or glyphosphate @ 2.01 kg/ha? is recommended. 3-4 rounds at 2-3 months intervals may be required to effectively control the weeds.

3. **Intercropping:** Rubber is planted at a wide spacing and hence sufficient land and light is available in the inter row areas during the initial years for intercropping. Intercrops should be selected based on slope of the land, light availability in the plantation and marketability. If the slope is less than 5 per cent, any intercrop can be cultivated. If the slope is more than 5 per cent, intercrops which require less soil disturbances should be selected. If the slope is more than 25 per cent, intercropping is not advised.

   Shade tolerant perennial crops viz., coffee, vanilla on *Gyricidea* standards, *Garcinea* and cocoa can be cultivated along with rubber without adversely affecting growth and yield of rubber. One row of these crops can be grown between two rows of rubber. Establishment of these crops will be better under partial shade and hence cultivating banana or providing shade by artificial means is beneficial. These crops should be manured separately. All other cultivation practices should be followed as per the recommendation for the respective crop. Yield of these crops will decrease after canopy closure. In mature rubber plantations, permitting more biodiversity through crops with canopy underneath rubber will improve soil moisture status during dry seasons and sustain soil nutrient status.

4. **Manuring:** Three stages of growth namely nursery, immature and mature can be recognised in the life of a rubber tree. The manuring differs according to the stages of growth. The Rubber Research Institute in India recommends the following manurial schedule:

(a) **Seedling nursery:** Incorporation of 25 kg of compost and 2.5 kg of rock phosphate once in three years per 100 sq. metre of the nursery bed is practised. Application of 25 kg of 10:10:4:1.5 NPKMg mixture per 100 m^2 of the nursery bed 6 to 8 weeks after planting and applicationof 12.5 kg of the mixture per 100 m^2 6-8 weeks after the first application but before mulching are followed.

(b) **Budwood nursery:** 1.5 kg of powdered rock phosphate per 100 m^2 of the nursery bed is applied as a basal dressing at the time of preparing the nursery bed. Besides 250 g of 10:10:4:1.5 NPKMg mixture per plant in two split doses of 125 g each, the first dose is applied two to three months after planting the budded stumps or cutting back, if budding is carried out *in-situ* and the second dose eight to nine months after planting.

(c) **Immature trees at pretapping stage:**

Year of planting	No. of application	Time of application	Dose of NPKMg mixture (10:10:4:1.5) per plant (g)
First	1	Sept-Oct	225
Second	2	April-May	450
		Sept-Oct	450
Third	2	April-May	450
		Sept-Oct	550
Fourth Year onwards till tapping	2	April-May	550
		Sept-Oct	550

(d) **Rubber trees under tapping:**

Application of 65 kg urea, 165 kg rock phosphate and 50 kg Muriate of Potash per hectare or 10-10-10 NPK mixture at the rate of 300 kg per hectare is recommended. In plantation where the trees show deficiency symptoms of magnesium (interveinal yellowing of leaves) addition of 10 kg commercial Magnesium sulphate to every 100 kg of the mixture is recommended during September-December.

Tapping

The rubber trees attain tappable stage in about seven years provided they possess the required girth of the trees. Seedling must attain a girth of 55 cm at a height of 50 cm from the ground. In the case of budded trees the girth should be 50 cm at a height of 125-150 cm from the bud union. Tapping is the periodical removing of thin slices of bark to extract rubber latex. Tapping is done by skilled men. While tapping the depth should be one mm close to cambium without any damage to it, otherwise callus formation will take place causing swellings.

Tapping has to be done on a slope of 30° to the horizontal zone in the case of budded trees and 25° in the case of seedlings (Fig. 16). Tapping is done early in the morning as late tapping will cause reduction in the flow of latex. In the early morning the turgor pressure in the latex vessels is high and rapid flow of latex occurs.

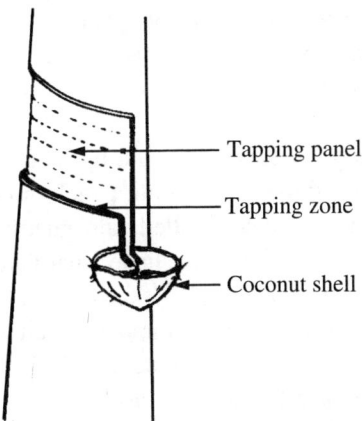

Figure 16   Tapping in rubber.

Tapping system: The following tapping systems are generally followed in India.

	Intensity	Remarks
S_2d_2 - Half spiral, tapping alternate days for 6 months and rest for 3 months.	100%	Recommended for budded plants
S_2d_3 - Half spiral, tapping at every three day for 6 months and rest for 3 months.	67%	Recommended for clonal seedlings.
S_2d_1 - Half spiral, daily tapping.	200%	Followed by small growers but it favours brown bast incidence and causes early deterioration of trees.

In South India, rubber trees shed their leaves during December-January and immediately again they put forth new leaves and flowering. During this period the trees are given rest since the yield of rubber will be poor if tapped. The yield of rubber steeply increases year by year and the peak is reached 14-18 years after planting. Then it slowly declilnes. After 40 years it may not be economical to maintain the trees. The latex yield will vary with the clone, age of the trees, fertility of the soil, climatic conditions and skill of the tapper. In the case of old trees, tapping may be done intensively adopting a system of two half spirals, one at normal basal level and the other at a higher level on the opposite side and away from first one atleast 120-180 cm.

Intensive tapping prior to felling of the old trees is called **slaughter tapping.** It is often done at higher levels, sometimes even on branches with the help of ladders and not on the usual renewed bark levels. As the objective of slaughter tapping is to extract as much latex as possible from the available bark, no consideration is given to the technique, intensity of standard of tapping.

Tapping is not done on rainy days but by fixing a polythene rain guard to the trunk of the tree above the tapping panel, tapping can be carried out during rainy season also. This is called rain guarding. Four types of rain guarding *viz.,* polythene skirt, tapping shade, guardian/kissan rain guards and tapping shield are followed. About 35-40 additional tappings per annum can be obtained by rain guarding the trees under the alternative-daily system. It is recommended in areas where the annual yield is 700 kg/ha or more and where normally more than 25 tapping days are lost by rain.

A number of chemicals have been shown to influence the flow of latex after tapping, among which, ethrel (2-chloroethylphosphonicacid) has been found to stimulate and increase the yield of rubber latex two fold. Ethrel has to be diluted with coconut oil to have ten per cent active ingredient and is applied thrice during the year i.e., March, August, September and November.

Yield of Rubber

In South India, the annual yield of rubber is about 375 kg per hectare per annum from the seedlings, whereas budded plantations yield 900 to 1000 kg of rubber/ha.

Processing of Rubber

Latex is a white or slightly yellowish opaque liquid with a specific gravity which varies between 0.974 and 0.986. It is a weak lyophilic colloidal system of spherical or pear shaped rubber globules suspended in an aqueous serum. The rubber globule is surrounded by a protective layer of proteins and phospholipids which impart the lyophilic colloidal nature to latex and the stability of latex is due to the negative charge present on the protective layer. Fresh latex, as it comes out from the tree is slightly alkaline or neutral. It becomes acidic rapidly due to bacterial action. The formation of organic acids neutralizes the negative charge on rubber particles and the latex gradually thickens and gets coagulated on keeping. Therefore, fresh latex cannot be kept for long without coagulation. The general composition of latex is rubber 30-40 % resins 1-2.0 %, protein 2-2.5 %, sugar 1-1.5 %, ash 0.7-0.9 % and water 55-65 %. The latex that flows out from the rubber trees on tapping is channeled into a container, generally coconut shell cups, attached to them. Latex collected in coconut shell cups is transferred to clean buckets, two to three hours after tapping. The latex which gets dried up on the tapping panel (tree lace) and the collection cups (shell scrap) also form a part of the crop and are collected by the tapper in baskets just prior to tapping. The latex spilt including overflown on the ground (earth scrap), when gets dried up, is also collected once in a month. Normally 10-20 per cent of the total crop constitutes the tree lace, shell scrap and earth scrap. Rubber can be processed and marketed as

1. **Preserved latex concentrates:** The latex is collected in the storage tank; from there it is brought to a centrifuge machine, rotating at 1440 rpm. Due to the centrifugal action, liquid portion comes out. The upper layer, the concentrated latex, is collected and brought to bulking tank and mixed with chemical and packed in drums. 60 % rubber is present in it. Skim latex is taken to another tank and sulphuric acid is added and coagulated and milled to get skim crepe. It is of poor quality while the concentrated latex fetches very higher price.

2. **Dry ribbed sheet rubber:**   Anti-coagulants (solutions of ammonia, formalin or sodium sulphite) are added to the cups to prevent the coagulation of latex before it reaches the factory. The latex so collected is bulked and then strained to remove the impurities. It is then diluted to a standard consistency of 12-15 % rubber. Special hydrometers like metrolac, latex meter are employed to measure the percentage of rubber. After dilution, the latex is strained through a 60 mesh screen for the second time. Then it is poured into the special coagulating tanks or aluminium pans which are divided into many compartments by thin aluminium sheets and acetic acid or formic acid is used for coagulation. Slow coagulation produces a soft rubber which is easy to work on the rollers. The acid is to be added quickly and mixed thoroughly with the latex. Froth formed is removed to avoid formation of bubbles on the surface of rubber sheets. After coagulation, rubber sheets are repeatedly washed several times with changes of water and passed through hand or power operated rollers. In the roller excess water and dissolved impurities are pressed and squeezed out. The surface of the rollers may be either smooth or grooved or zig zag or straight or diamond pattern, its impression is normally left on the surface of the sheets when they come out of the press. These sheets are hung in shade for two to three hours for dripping in a dust free place. They are then taken to smoke houses for thorough drying. Smoking of rubber sheets is done to dry the sheets properly and to avoid formation of blisters. In the smoke house, the sheets are smoked at a low temperature of 48-50°C with fairly high humidity during the first day subsequently during 2nd to 4th day the temperature being 68°C with low relative humidity. They are taken out, graded and packed. Such products are known as smoked sheets or dry ribbed sheet rubber. Various grades of rubber sheets are **RMAIX**, **RMA**-1, **RMA**-2, **RMA**-3, **RMA**-4 and **RMA**-5. High grade rubber sheets are clear, free from blisters, translucent and of a golden brown colour and fetch a better price.

3. **Dry crepe rubber:**   When coagulum from latex or any form of field coagulum after necessary preliminary treatments is passed through a set of creping machines to get crinkly, lace-like rubber called 'crepe rubber' after drying. Various grades of crepe rubbers are EPC super 1X, EPC1X, EPC2X, and EPC3X.

Rubber wood: For a long time, rubber wood (after felling the old trees) was considered fit only for burning as it is susceptible to insects and fungal attack. But now, after proper chemical treatment, rubber wood provides enough strength and durability of any semi- hard wood available in India and can be used for the manufacture of useful articles like door and window components, furniture, wall paneling, interior decoration, tool handles etc.

Plant protection: The important diseases and pests of rubber and their control measures are furnished below:

Disease/Pest	Symptoms	Control measures
Abnormal leaf fall (*Phytophthora palmivora*)	Infected leaves fall in large numbers prematurely.	Spray Bordeaux mixture (1%) as prophylactic measure, prior to the onset of South West monsoon.
Powdery mildew (*Oidium heveae*)	Ashy coating noticed on tender leaves	Dusting 3 to 6 rounds at 10-15 days interval using 11-14 kg of 325 mesh fine sulphur dust per round per hectare.
Scale insects (*Saissetia nigra*)	Severely affected portion dry up and die.	Spray malathion at 0.05 % concentration.
Mealy bug (*Perrisiana virgata*)	Severely affected portion dry up and die.	Spray profenophos 50 EC @ 2 ml/L

CHAPTER

12

Cocoa

COCOA (*Theobroma cacao* L.)

Family: **Malvaceae**

Cocoa, (*Theobroma cacao*) is a beverage crop introduced for cultivation in India in the early 1965. It is native of Amazon valley of South America and now cultivated largely in Ghana, Nigeria, Sierra Leone, Cameroon, Brazil, Ecuador, West Indies and Malaysia. In India, the current area is estimated to be 46,318 hectares with production of 11,144 MT. However, our annual requirement is about 30,000 MT which is predicted to rise. The national productivity is 550 kg dry beans per ha. Kerala leads in production with an area of 11,044 hectares contributing 6344 MT of cocoa beans with a productivity of 592 kg per hectare. Tamil Nadu occupies third in cocoa cultivation and the area reported under this crop is 15,000 ha with an annual production of 350 MT.

Cocoa is small (4 to 8 m height) evergreen perennial tree. Vertical growth of the seedlings (chupon) terminates at jorquette, where four or five branches develop. Flowers are borne on thickened leaf axils (cushions) on trunk and matured stem (cauliflorous) from second year of planting. It is an ever flowering plant and is highly floriferous. But flowering is not uniform throughout the year; peaks of flowering are some months of the year. Depending upon the pattern of rainfall and prevailing temperature, the peak flowering varies in different areas of cocoa cultivation.

170

Each cushion will bear up to 50 flowers in one season. Although more flowers are produced in each cushion, only 1-5 per cent of the flowers are successfully pollinated to produce a pod. Almost 60 % of the flowers produced by the cocoa plant are not pollinated and drop after 48 hours. Important group of pollinating insects are midges, belonging to the several genera of the family *Ceratopogonidae* besides many other insects including ants, aphids, fruit flies and thrips. The mechanism of self incompatibility (due to the inhibition of growth of the pollen tubes on its own stigma) makes cross pollination as the rule.

Individual flowers are pedicellate, pentamerous, regular and hermaphrodite. Outer most whorl comprises of calyx with five sepals and polysepalous in nature. They are greenish to light cream in colour, polypetalous corolla with five petals. Each petal consists of pouch like lower part and a spatula shaped upper part. These two portions are attached with the help of a thin connective. Two purple coloured guidelines can be seen on the petal. Androecium consists of 10 stamens arranged in two whorls, outer five represent aborted stamens/staminodes. They are purple in colour and form a ring around the style. Inner five are fertile in nature. Filament of fertile stamen bends outside and the anther lobes are concealed inside the pouch like portion of petal. Gynoecium is with five carpels, syncarpous in nature. Ovary is superior with short style and pentafid stigma. Number of ovules ranges from 40-60 which are arranged in axile placentation at base and parietal above.

Varieties

Commercial cocoa has two major varieties, Criollo and Forestero which differ in many aspects as follows:

	Character	Criollo	Forestero
1.	Cotyledons	Plumpy and white when fresh and turn cinnomon coloured on fermentation.	Flat and purple when fresh and turn dark chocolate brown on fermentation.
2.	Pod colour	Dark red	Yellow
3.	Other pod characters	Rough surface, ridges prominent, pronounced point and thin walled.	Smooth, inconspicuous ridges, thick walled, melon shaped with rounded end.
4.	Flavour and aroma	Bland flavour	Harsh flavour, bitter taste
5.	Duration of fermentation	3 days	6 days

| 6. | Adaptability in India | Poor adaptability and less yield potential and hence discouraged for commercial cultivation. | Good adaptability and high yielding and hence recommended for commercial cultivation. |

Other types of cocoa include (1) Trinitario from Trinidad which is said to be a hybrid between Criollo and Forestero with highly variable pod characters, (2) Amelonado, a Forestero type bean with a melon shaped pod, cultivated in West Africa and (3) Amazon, a relatively new type collected from the Amazon forests which has got vigour and high yield.

In India, KAU, Thrissur and CPCRI, Regional Research Station, Vittal have developed some improved cultivars which are briefly described hereunder:

S. No	Name of the Clone	Pod Traits	Yield (No. of Pods/ tree)
1	CCRP –1	Mature pods weigh 385 g with 46 beans and 0.8 g dry bean weight	56
2	CCRP – 2	Mature pods weigh 311.3 g with 45.5 beans and 0.9 g dry bean weight.	56
3	CCRP – 3	Mature pods weigh 240.6 g with 42.3 beans and 1.0 g dry bean weight.	68
4	CCRP – 4	Mature pods weigh 402 g with 45 beans and 1.1 g dry bean weight.	68
5	CCRP – 5	Mature pods weigh 425 g with 42 beans and 0.8 g dry bean weight.	38-55
6	CCRP – 6	Mature pods weigh 895 g with 48 beans and 1.9 g dry bean weight.	55
7	CCRP – 7	Mature pods weigh 526 g with 42 beans and 0.9 g oven dry bean weight.	78
8	CCRP – 8	Hybrid between CCRP 1 × CCRP 7, mature pods 389 g with 49 beans and 0.88 g dry bean weight.	90
9	CCRP – 9	Hybrid between CCRP 1 × CCRP 4, mature pods 370 g with 37 beans and 0.8 g dry bean weight.	106
10	CCRP – 10	Hybrid between CCRP 3 × GVI 68, mature pods 332 g with 41 beans and 1.1 g dry bean weight.	80

11	VTLCC 1	Clonal selection from Nigerian type, pod weighs around 320 g with 36 beans and the dry bean weight is 1.05 g. Fat content is 52.5 %.	75
12	VTLCH 1	It is a hybrid of I-56 × II-67, each pod weighs around 350 g, with 42 beans and single dry bean weight is 1.00 g. Fat content is 53.6 %.	50
13	VTLCH 2	A hybrid of ICS-6 × SCA-6, each pod weighs around 350 g, with 40 beans and single dry bean weight is 1.15 g. Fat content is 54 %.	70
14	VTLCH 3	A hybrid of I-67 × NC- 29/66, each pod weighs around 430 g with 43 beans and single dry bean weight is 1.07 g. Fat content is 52 %.	45
15	VTLCH 4	A hybrid of I-67 × NC- 42/94, each pod weighs around 425 g, with 43 beans and single dry bean weight is 1.01 g. Fat content is 53 %.	40

Note: 1-10 were developed from from KAU, Kerala while the rest were developed from CPCRI, Regional Research Station, Vittal, S.No 1-7 and 11 are clones and the rest are hybrids.

Climate and Soil

Cocoa is a crop of humid tropics requiring well distributed rainfall. A minimum of 90 to 100 mm rainfall per month with an annual precipitation of 1500-2000 mm is ideal. However, it can also be grown in other regions by supplementing rainfall with irrigation during dry periods. However, for successful cultivation the dry months should not exceed 3 to 4 months. This limits the distributions of cocoa to within 20° latitude on either side of the equator. Cocoa tolerates a minimum temperature of 15°C and a maximum of 40°C, but temperature around 25°C is considered as optimum. It can be grown in place from sea level up to an elevation of 1000 m, more ideal is up to 500 m above sea level.

Cocoa grows on a wide range of soils but loose soils which allow root penetration and movement of air and moisture are ideal. It should retain moisture in the soil during dry season as cocoa requires regular supply of moisture for proper growth. Though cocoa can be grown in soils with pH range from 4.5-7.0, it thrives better in neutral soil.

Planting Material

Cocoa can be propagated from seeds or vegetatively from buds, grafts and cuttings. Seed pods may be collected from trees yielding 80 or more pods per year with pod weight 350-400 g. Seeds of cocoa are recalcitrant without dormancy period. Seeds once extracted from pods should be sown immediately as they loose viability quickly. Germination can be extended for some more days by storing freshly extracted seeds in moist charcoal and packed in poly bag for a period of four weeks. Seeds mixed with sawdust, testa removed and treated with fungicide either by washing in a solution or by dusting can be stored to preserve its viability for three to four weeks. Seeds normally start germinating in about a week and will continue for another one week. Seeds should be kept horizontally or vertically with hilum end down and just covered with sand. Pushing of seeds deeply into the potting mixture should be avoided because lack of air may affect seed germination since it exhibits epigeal type of germination where cotyledons are taken above ground in the process. This stage of germination is called as soldier phase. Healthy seeds from well matured pods usually give a germination of 90 to 95 per cent.

Before sowing, the seeds are rubbed with dry sand or wood ash to remove mucilage. The beans are planted with their pointed end upwards, either in plastic bags (25 × 15 cm size, 150 gauges) or in raised beds. If sown in beds, young seedlings are usually transplanted into polythene bags after about two weeks of germination. The seedlings are ready for transplantation to the field after about 3 to 4 months or they attain a height of 30 cm.

As the seedling progenies show wider genetic variability due to its heterozygous and cross pollination condition, asexual or vegetative propagation is followed to maintain true to types. Grafting and budding are being followed in multiplication of cocoa. It also ensures multiplication of identified high yielding clones in large quantities. Though vegetative propagation of cocoa by budding, rooting of cutting and grafting are feasible, the widely accepted methods in India are budding and grafting.

Vegetative propagation	Rootstock	Scions	Success (%)
Budding: Patch budding	10-12 month old seedlings	Bud patch of 2.5 cm length and 0.5 cm width from the bud wood	85
Grafting: Soft wood grafting	3-4 month old seedlings	Scion stick of 12- 15 cm length with 2- 3 buds	70

Establishing Plantation

Cocoa, whose natural environment is the lower storeys of the forest, requires shade when young and also to a lesser extent when grown up. Young cocoa plants grow best with 50 percent full sunlight. Therefore, it can be grown well in the partially shaded conditions prevailing in the arecanut and coconut gardens in our country or as a pure plantation in forest lands by thinning and regulating the shade suitably. It is planted at a distance of 2.5-3.0 m both between and within rows, either in the beginning of the monsoon, in May-June or at the end of the south west monsoon, in September.

Cocoa under arecanuts and coconuts is the cropping systems which can be adopted advantageously in Kerala, Karnataka and Tamil Nadu. In arecanut gardens where the spacing is 2.7 × 2.7 m cocoa is interplanted in alternate rows at a spacing of 5.4 × 2.7 m (686 cocoa plants/ ha). In coconut gardens, it can be planted 2.7 m apart in a single row besides planting one cocoa in between two coconut plants (500 plants/ha). Under the double hedge system, cocoa is planted in two rows adopting a spacing of 2.7 m within the row and 2.5 m between rows of coconut planted at a normal spacing of 7.5 × 7.5 m and above (Figure 17). In oil palm plantations raised at 4.5 × 4.5 m spacing, it can be planted in such way that Five cocoa plants would come between four oil plants resulting in 400 plants per hectare.

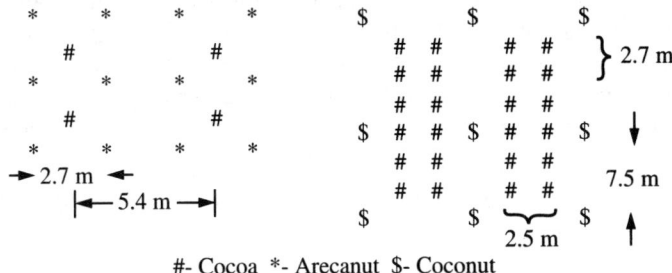

#- Cocoa *- Arecanut $- Coconut

Figure 17   Cocoa as intercrops in arecanuts and coconut gardens.

Mulching is recommended for good establishment especially under rainfed situation. The circle (60-90 cm across) around each seedling may be cleared and applied with leafy mulch up to 30 cm thick. Additional mulch must be added every 3 months.

Manuring

An annual application of 100 g N, 40 g P_2O_5 and 140 g K_2O per tree per year in two split doses is recommended. During the first year of planting, the plants may be given one third of the above dose, while in the second and third year and above, two third and full doses of fertilizers are applied respectively. The fertilizer is applied in two splits, the first dose in March-April and the second dose in September-October. Fertilisers may be applied uniformly around the base of the tree up to a radius of 75 cm and forked and incorporated into the soil. Drip fertigation is done at weekly intervals at different stages of crop growth helps to increase the yield more than 50 percent.

Micronutrient deficiency especially zinc is very common in South India. Symptoms are chlorosis of the leaves which appear in patches and in advanced stages the green areas are found only along the vein margins, giving a vein banding appearance to the leaves. Affected leaves show mottling and crinkling with wavy margin. Most of the younger leaves become narrow and sickle shaped showing characteristic 'little leaf' symptom. Symptoms on twigs include rosette and dieback. Shortening of internodes causes a rosette type of growth. Management strategy includes foliar spray of a mixture of 0.3 % (3 g in 10 litre of water) Zinc sulphate and 0.15 % (w/v) lime.

Irrigation

Cocoa plants require continuous supply of moisture for optimum growth and yield. During summer, the plants will have to be irrigated at weekly intervels. If adequate water supply is not ensured in summer months, the yield will be reduced and under mixed cropping systems, if there is severe drought, the yield of both the crops may be affected. Drip irrigation is folowed in Tamil Nadu and Andhra Pradesh where in the 1st year, 2nd year and 3rd year planting, water @ 3-5, 10 and 20-25 litres/plant/day are normally be provided.

Pruning

Cocoa plants are grown under the shade of arecanut and coconut plantations. It is therefore necessary to regulate the canopy size and shape of plants so that the main crop is not affected. Proper and systematic pruning is essential in cocoa cultivation. Different types of pruning are generally followed.

Initally, **'formation pruning'** is done in young plants, mainly to adjust the height of first jorquette. The jorquette is allowed to form at a height of 1.5 – 2.0 meters that will help in easy cultural operations. This pruning will decide the number of jorquette and height of first jorquette.

Pruning in mature cocoa includes two types viz., sanitary pruning and structural pruning. In **sanitary pruning,** all unnecessary chupons, diseased or unnecessary branches and water shoots including damaged pods and over ripe pods epiphytes, climbing plants, ant nests etc, are removed to maintain the health and vigour of the tree.

Structural pruning is done to shape the canopy to desired size and architecture. For optimum productivity, proper canopy management to maintain shape and size is required.

In either case care must to taken, when removing large branches to ensure that exposed wood surface is not damaged, to prevent the entry of fungi or insects, which is achieved by applying fungicides (Bordeux paste) immediately after the pruning.

Pruning also depends upon the type of planting materials used; in the case of 'seedling material' first jorquette between 1 to 2 meters and 3-4 fan branches are retained with vertical height restricted to first jorquette. It is mainly to shape the canopy for desired shape, which should be umbrella –shaped. The canopy spread of 3.8 to 4.0 m and height 2.7 m are the ideal canopy architecture for optimum yield. In the case of graft / budded plants, primary pruning should be done to obtain a supporting framework of one or more upward growing main stems. Then drooping or inward growing branches are to be removed. Secondary pruning is suggested to develop well- shaped canopy and desired canopy should be maintained in umbrella shaped form with about 3.8 m to 4.0 m spread and 2.7 m height depending upon the space and main crop in which cocoa is under planted and grown.

It is important to note that the maximum leaf area should be maintained with pruning practices to avoid self – shading. Pruning is usually done annually in July- August. Under irrigated field of Tamil Nadu pruning twice following harvesting is recommended. The proper pruning of cocoa ensures adequate ventilation in garden; maintain tree height, makes spraying and harvesting operations easier. When cultivation as mixed crop under palms a maximum of two–storey canopy architecture may be maintained.

Harvesting

Cocoa flowers from the second year of planting and the pods take about 140 to 160 days to mature and ripen. Each pod will have 25 to 45 beans embedded in white pulp (mucilage). Generally cocoa gives two main crops in a year i.e., September-January and April-June, off-season crops may be seen almost all through the year, especially under irrigated condition. The shriveling and mummifying of some young fruits is a common phenomenon (**Cherelle wilt**) in all cocoa gardens. In the early stages, the fruits lose their lusture and in four to seven days, the fruits shrivel. The fruits may wilt but do not abscise. Many factors are involved in the causation of this malady.

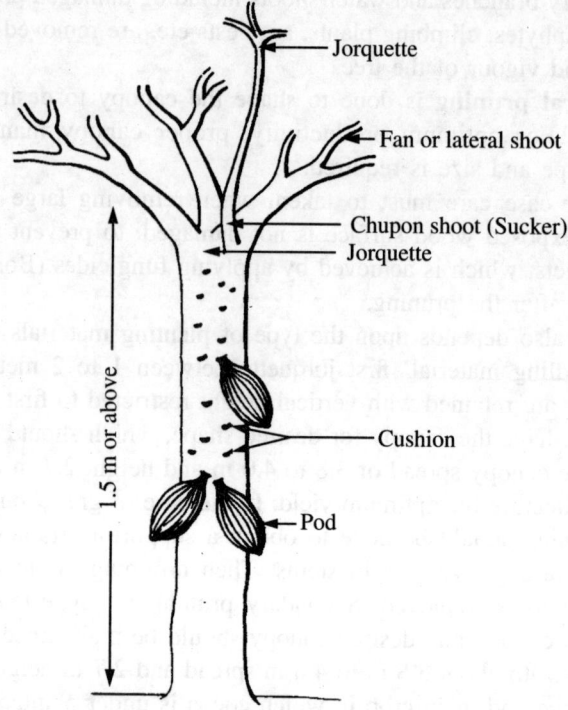

Figure 18 Properly trained cocoa tree

The most important factors are insects, fungi, nutrient competition, over production etc. Hence, the remedial measures will depend upon the nature of the causative factors involved.

Only ripe pods have to be harvested without damaging the flower cushion, at regular intervals of 10 to 15 days. The pods are harvested by cutting the stalk with the help of a knife. The harvested pods should be kept for a minimum period of two to three days before opening for fermentation. For breaking the pods crosswise, wooden billet may be used and the placenta should be removed together with husk and the beans are collected for fermentation. A pod will have about 30 to 45 seeds covered with pulp or mucilage.

Processing

1. **Fermentation:** The beans should be fermented to develop chocolate flavour, reduce bitterness, loose its viability, remove the mucilage coating and enable the cotyledons to expand. Fermentation is done

immediately after collecting the beans from the pods. There are two popular methods of fermentation using either trays or boxes.

2. **Box method:** In this traditional method, boxes of various shapes and sizes are used. The smallest one has the measurements of 60 cm × 60 cm and will hold about 150 kg wet beans. The bottom of the box has a number of holes of 1 cm diameter spaced at about 10 cm apart. Three such boxes are arranged in a row so that beans can be transferred from one box to the other. The beans are placed in the top most and covered with banana leaves or gunny bags. After 2 days, the bean should be uncovered and transferred into the second box and then to the third box after another 2 days. On the sixth day, fermentation is completed and beans can be taken out for drying.

3. **Tray method:** This method is used only for fermenting forestero cocoa beans. The normal size of the tray is 90 × 60 × 12 cm with a capacity to hold about 40 kg wet beans. The bottom of the tray is either slotted or drilled to make 1 cm holes on a 4 cm sq. pattern. A minimum of 4 trays are needed for successful fermentation. All the trays are filled with beans. The top-most tray is covered with banana leaves or sacks. The fermentation is faster here and is completed in about 4 to 5 days. This method is more convenient for large growers as the trays can be easily handled and no mixing is required and the period of fermentation is less.

4. **Basket method:** Bamboo or cane basket of suitable size having one or two layers of banana leaves at bottom to drain the sweating is filled with the beans and the surface is covered with banana leaves. After one day the basket is covered with thick gunny sacks. The beans are mixed thoroughly on the third and fifth days and covered with gunny sacks. The fermentation will be completed at the end of the sixth day and the beans are withdrawn for drying.

5. **Drying:** After the fermentation, the beans can be dried by sun drying or artificial drying as the fermented cocoa beans have considerable moisture (55 %). Sun drying is good as it gives superior quality produce when compared to artificial drying. The fermented beans are spread in a thin layer over a bamboo mat or cement floor and dried for 5 to 6 days. The beans are to be stirred from the time to time for uniform drying. The moisture content of well dried beans is around 6 to 7 per cent.

During the monsoon period, artificial drying has to be adopted. Electric ovens or conventional Samoan type drier could be used.

The duration of artificial drying varies from 48 to 72 hours at 60 to 70°C. The drying of beans at high temperature should be avoided as it results in low quality end product. Slow drying in the initial stage has given better quality beans. Well dried beans when shaken should give a metallic sound.

Cocoa quality depends on various factors, but primarily on the cocoa variety and the post-harvest handling. Generally, fine or flavour cocoa beans are produced from Criollo or Trinitario varieties, while bulk cocoa beans come from Forastero trees. The table below shows the relationship between colour, degree of fermentation and flavour

Colour of beans	Degree of fermentation	Flavour of roasting
Brown	Fully fermented	Strong cocoa flavour, balance of acidity, astringency and bitterness
Brown/ purple	Partly fermented	Good cocoa flavour, higher acidity, astringency and bitterness
Purple	Low fermentation	Low cocoa flavour, strong acidity, astringency and bitterness
Greyish or Black	Unfermented	Absence of cocoa flavour, predominantly acid, astringent and bitter. Over all sour flavour

6. **Grading and Storage:** The flat, slaty, shriveled, broken and other extraneous materials are removed. The cleaned beans are packed in fresh polythene-lined (150-200 gauge) gunny bags. The bags are kept on a raised platform of wooden planks. The beans should not be stored in rooms where spices, pesticides and fertilizers are stored as they may absorb the odour from these materials.

The roasted product of the dried beans is called as cocoa which is used for the manufacture of various products. When cacao nibs are ground, the resulting product is called chocolate liquor or mass. The fat that is pressed from chocolate liquor is termed as cocoa butter. It is mainly used for the manufacture of chocolates, in pharmaceutical preparations and soap making.

Plant protection: Many pests and diseases are known to infect cocoa and cause economic losses.

	Pest/diseases	Symptoms	Control measures
1.	Mealy bugs *(Planococcus lilacinus)*	Adult and young ones suck tender shoots, cushions, flowers, cherelles and pods.	Spot application of Profenophos 200 ml in 100 litres of water.
	Stem borer *(Zeuzera coffeae)*	Caterpillars bore into the branches and trunks.	Prune and destroy the affected branches, apply carbaryl paste.
3.	Aphids *(Toxoptera aurantii)*	Adults and nymphs feed on young leaves, succulent stem, flowers and small cherelles.	Spray dimethoate 1.5 ml in one litre water.
4.	Rodents-Rats and squirrels	Pods are damaged	Keep 10 g of bromadiolone (0.005 %) wax cakes or mix zinc phosphide with dried fish on the branches for rats, keep mechanical traps for squirrels.
5.	Blackpod disease *(Phytophthora palmivora)*	Pods turn chocolate brown to black, beans discoloured.	Remove infected pods at frequent interval; spray 1 % Bordeaux mixture during monsoon twice at 45 days interval. Application of *Pseudomonas fluorescens* (Pf1) liquid formulation @ 0.5 % by soil and foliar spray 3 times per year (June, October and February) is also effective.
6.	Canker *(P. palmivora)*	Brownish water soaked lesions on the outer bark, which then turn into rusty deposits.	Remove the infected tissues and apply Bordeaux paste.
7	Vascular streak die-back *(Oncobasidium theobromae)*	Characteristic yellowing of one or two leaves on the second or third flush behind the growing tip, diseased leaves fall off within few	Disposal of diseased branches, and regular pruning of chupons on the trunk.

		days and entire leaves subsequently show similar symptoms. Infected shoot if split longitudinally, shows brown streaking of woody tissues extending well beyond the region of yellow leaves.	

13

Cashew

CASHEW (*Anacardium occidentale* L.)

Family: **Anacardiaceae**

Cashew is native of South Eastern Brazil, from where it was introduced to Malabar coast of India in the sixteenth century to cover bare hills and for soil conservation. It gained commercial importance in 1920. Cashew is now grown in India, Brazil, Vietnam, Tanzania, Mozambique, Indonesia, Sri Lanka and other tropical Asian and African Countries. The world production of Cashew is estimated to be around 20.8 lakh tons. According to a recent statistics, India grows cashew in about 9,23,000 ha and produces 6,13,000 MT of nut with an average productivity of 695 kg/ha Cashew is grown mainly in Maharashtra, Goa, Karnataka and Kerala along the west coast and Tamil Nadu, Andhra Pradesh, Orissa and West Bengal along the east coast. To a limited extent it is grown in Manipur, Meghalaya, Tripura, Andaman and Nicobar Islands and Chhattisgarh and also in plans of Karnataka. At present Maharashtra is ranking first in area, production and productivity in the country.

African countries produce large quantities of cashewnut, but they are not processed into consumable products because of difficulties in organising the native labour. India imports raw cashew nuts which are processed and converted into cashew kernels and cashew shell liquid and exported to the countries like USA, Canada, United Kingdom, Australia and Russia. India has been exporting cashew kernels since early part of 20th century. Over the years, both the export earnings (about Rs. 2700 crores) as well as quantity of kernels have been increasing. The established processing capacity of raw nuts is around 12 lakh tons. However, domestic production of raw nuts is around 5.7 lakh tons only. Thus, presently India is importing raw nuts from African and other countries to a tune of 5.8 lakh tons to meet the demand of cashew processing industries. Cashew industry provides employment to 5.5 lakhs people mainly to women. Recently, African countries have started processing their own raw cashews and hence to keep the processing industries viable and to meet the increasing demand for processed nuts from global market, India is contemplating to increase its production and productivity.

Botany

The cashew tree is a low spreading, evergreen tree with a number of primary and secondary branches and with a very prominent tap root and a well developed and extensive network of lateral and sinker roots. The leaves are alternate, simple, glabrous, obovate, round and pinnately veined. The inflorescence is an indeterminate panicle of polygamo monoecious type i.e. flowers is either bisexual or staminate but both occur intermixed in the same inflorescence. The proportion of the perfect flowers varies from as low as 0.45 to 24.9 percent in different trees. On the same tree, the perfect flowers are larger in size than the staminate. Pollination is carried out by flies, bees and ants as well as by wind. Under normal conditions, nearly 85 percent of the flowers are fertilized of which only 4 to 6 percent reaches maturity, the remaining being shed away at various stages of development.

The fleshy peduncle, the 'cashew apple', is juicy and sweet when ripe. The apple varies in size, colour, juice content and taste. It is a rich source of vitamin C and sugar. Cashew apple is highly nutritious and is a valuable source of sugars, minerals and vitamins. Traditionally, several products are prepared from cashew apple, including those with medicinal properties, some of which are still used. However, it is not commercially exploited anywhere in the country except in Goa where it is used for the preparation of cashew feni. Kerala Agricultural University has developed many products from cashew apple such as Cashew apple syrup and drink, Cashew Apple-Mango mixed jam, Cashew Apple Candy, Cashew Apple Pickle etc. The residual waste after the above products can be utilized for the manufacture vermi compost, animal feeds and alcohol.

The cashew fruit is a kidney shaped drupaceous nut, greenish grey in colour. The nuts vary in size, shape, weight (3 to 20 g) and shelling percentage (15-30 percentage).

Based on the nature of branching in cashew, they are described either as 'intensive' or 'extensive'. The intensive shoot grows to a length of about 25 to 30 cm and terminates in a panicle. Simultaneously, three to eight laterals arise within 10 to 15 cm of the apex. Some of these laterals shoots may also terminate in panicles in the same flowering season, repeating the same growth pattern, giving a well covered bushy appearance to them. In the extensive type, the shoot grows to a length of 20 to 30 cm and rests. A bud sprouting 5 to 8 cm below the apex leads further growth. This process of growth continues for two or three years without any flowering. This kind of growth pattern produces a spreading tree. Even though both kinds of branching are observed in all trees, one type dominates in a tree. High yielding trees have more than 60 percent intensive branches, whereas low yielders have less than 20 percent.

Climate and Soil

It is a hardy tropical plant and does not exact a very specific climate. It can come up in places situated within 35° latitude on either side of the equator and also in the hill ranges up to 700 m MSL. It can grow well in places receiving rainfall from 50 cm to 250 cm and tolerate a temperature range of 25° to 49°C. It requires a bright weather and does not tolerate excessive shade.

Cashew is cultivated on a wide variety of soils in India like laterite, red and coastal sandy soil. To a limited extent, it is also grown on black soils. It can be also grown in hill slopes in virgin organic matter rich soils. They do not prefer water logged or saline soils.

Varieties

Since cashew is a highly cross-pollinated crop, planting of seedlings is not recommended now. Various cashew research centres have released improved clones which are either selections from seedling population or hybrids. They are briefly described Table-3.

Seed propagation: Seed propagation is seldom practiced now except to raise the rootstock materials. Seeds should be collected during the month of March to May and the heavy seed nuts which sink in water are alone mixed with 2 parts of fine sand. They take normally 15 to 20 days for germination.

Vegetative propagation: One year old shoots as well as current season shoots are used for air-layering. Though there is good root development in this method, heavy mortality occurs both at nursery stage and on the main field. Other drawbacks of this method are that it is cumbersome, time consuming and production of a limited number of layers per tree and such layers are also not suitable for cyclone prone areas as they do not have a tap root providing less anchorage to soil.

The vegetative propagation through cuttings is seldom practiced as success is very less. Similarly veneer grafting, side grafting and patch budding are also reported to be successful but the nursery period is quite long, 3 to 4 years. Recently 'epicotyl grafting' and 'soft wood grafting' are recommended for commercial scale adoption. In the case of epicotyl grafting, tender seedlings with height of 15 cm are selected as root stocks and a 'V' shaped cut is made after beheading it at a height of 4 to 6 cm from the cotyledons connective. The precured scion is collected and a wedge is made at the base of it, so as to exactly fit in the cut made at the base of the stock. The scion is exactly fitted in the stock and tied with polythene strips. The success of epicotyl grafting varies from 50 to 60 percent and depends upon high humidity, temperature, freedom from fungal disease, number of rainy days and rate of cambial growth. When the above method is adopted in 30 to 40 days old seedling, it is known as soft wood grafting. The success varies from 40 to 50 percent.

Planting

Pits of 45 × 45 × 45 cm are dug and filled with a mixture of top soil, 10 kg of farm yard manure and one kg of neem cake at a distance of 7m × 7m either way during June July and planted. In the case of seedling, 45 days old seedlings are transplanted. High density planting is now recommended. Under Tamil Nadu conditions, a spacing of 5 × 4 m accommodating 500 Plants/Ha is recommended under HDP while a spacing of 8 × 4 or 4 × 4 m is good under Kerala situation. Under irrigated situation, ultra high density planting (4 × 2 or 3 × 2 m spacing) can be followed under intensive management situation to obtain maximum yield.

After Cultivation

The interspaces may be ploughed after the receipt of the rain and intercrops like groundnut or pulses or tapioca raised normally till they reach their bearing stage. Although it is grown as a rainfed crop, Irrigation once in a weak from flushing to fruit maturity stage is good to increase the yield.

Table 3 Improved verities in Cashew

Name of the variety	Parentage	Yield (kg/tree)	Nut weight(g)	Shelling percentage	Remarks
A. Varieties released for cultivation in Andhra Pradesh					
Cashew Research Station, Bapatla					
BPP1	Hybrid tree No. 1 × T.No.273	17 (25)	5	27.5	13.2 % perfect flowers, apple medium sized,' yellow, 68 % juice content, 8 fruits/panicle. 7 % perfect flowers, 8 to 10 fruits per panicle
BPP2	Hybrid tree No. 1 × T.No.273	19 (25)	4	26.0	7 % perfect flowers, 8 to 10 fruits per panicle
BPP3	Clonal selection from germplasm type	16 (25)	6	28.0	–
BPP4	Clonal selection from germplasm type	12.5 (17)	6	23.0	Younger leaves with light pigmentation
BPP5	Clonal selection from T.No.1 (A.P)	42 (50)	5.2	24.0	–
BPP6	Clonal selection from T.No.56 (A.P)	42 (50)	5.2	24.0	–
BPP 8	Hybrid tree No. 1 × T.No.273	14.5	8.2	29.0	–

B. Varieties released for cultivation in Maharashtra

Vengurla-1	Clonal selection from germplasm (Vengurla)	23 (28)	6.0	31.0	8 % perfect flowers, intensive branching and compact type
Vengurla-2	Clonal selection from Germplasm (West Bengal)	24 (20)	4.0	32.0	8 % perfect flowers, short flowering and fruiting phase
VengurIa-3	Hybrid of Ansur 1 × Vetore-56	17	9.0	27.0	25 % perfect flowers, 7 fruits per panicle
Vengurla-4	Midnapore Red × Vetore-56	23	8.0	31.0	35 % perfect flowers, 6 fruits per panicle
Vengurla-5	Ansur Early × Mysore Kotekar	21 (14)	4.5	30.0	50.5 % perfect flowers 3-4 fruits/panicle.
Vengurla-6	Vetore 56 X Ansur-1	13.80	8.00	28.0	
Vengurla-7	Vengurla –3 X M10/4	18.50	10.00	30.5	
Vengurla-8	Vengurla-4 X M10/4	15.70	11.50	28.0	

C. Varieties released for cultivation in Tamil Nadu

Regional Research Station, Viruthachalam

VRI 1	Clonal selection from germplasm (S.Arcort)	7.4 (17)	5	20	14.6 % perfect flowers, 5 to 7 fruits/ panicle, intensive branching type
VRI 2	Clonal selection from germplasm (Chengleput)	6.0 (17)	5	28	10 % perfect flowers, 5-8 fruits/ panicle, intensive type of branching
VRI 3	A clonal selection from seedling tree.	10.0	7.2	29.1	Nut size is 7.18 g, kernel weight is 2.16 g, shelling percentage 29.1 % with export grade (w 210)

VRI 4	Selection from M44/3	7.2	6.30	29.1	Extensive branching, bisexual flower 12.4 %, Nut size is 6.30 g, kernel weight is 2.05 g, grade (w 240)
VRI(CW)-H1	M 26/2 (VRI-3) x M 26/1	13.2	7.2	30.05	Higher percentage of bisexual flowers (12.5 %),cluster bearing nature (6-10 fruits / panicle),Bolder nut (7.2 g), Larger kernel (W 210) and easy to peel the testa

D. Varieties released for cultivation in Karnataka

Agricultural Research Station, Ullal

Ullal 1	Clonal selection from germplasm type (Kerala)	19	7	31	2.3 % perfect flowers
Ullal 2	Clonal selection from germplasm type (Guntur)	18	6	30	7 % perfect flowers late flowering type, short harvesting duration
Ullal 3	5/37 Manjeri	14.7	7.0	30.7	--
Ullal 4	2/77 Tuni	9.5	7.2	31.0	--
Ullal 5	2/27 Nileshwar	10.5	9.0	32.8	--

Directorate of Cashew Research, Puttur, Karnataka

NRCC 1	3/28 Simhachalam- AP	10.00	7.60	28.80	--
NRCC 2	2/9 Dicherla	9.00	9.20	28.60	--

Agricultural Research Station, Chintamani, Karnataka

Chintamani-1	8/46 Taliparamba	7.20	6.90	31.00	--

E. Varieties raleased for cultivation in Kerala

Cashew Research Station, Anakkayam.

Variety	Parentage/Description				Vigorous type, intensive branching
Anakkayam 1	Selection from O.P of T.No.139 (A.P)	12	5.95	28	
Sulabha	A selection from op seedlings.	21.90	9.80	29.40	--
Mridhula	PTR 1-1				--
Cashew Research Station, Madakkathara					
Madakkathara-1	A selection from Open pollinated Seedling	13.80	6.20	26.80	--
Madakkathara-2	A selection from Open pollinated Seedling	17.00	7.25	26.00	--
Dhana	Hybrid (ALGD-1 X K-30-1)	10.66	8.20	29.80	--
Kanaka	Hybrid (BLA 139-1 X H3-13)	12.80	6.80	30.58	--
Priyanka	Hybrid (BLA 139-1 X K-30-1)	17.03	10.80	26.57	--
Dharashree	Hybrid (T30 X Brazil-18)	15.02	7.80	30.50	--
Amrutha	Hybrid (BLA 139-1 X H3-13)	18.35	7.18	31.58	--
Damodar	Hybrid (BLA 139-1 X H3-13)	13.65	8.20	27.27	--

Raghav	Hybrid (ALGD-1-1 X K-30-1)	14.65	9.20	26.60	--
Akshaya	Hybrid (H-4-7 X K-30-1)	11.78	11.00	28.36	--
Anagha	Hybrid (T 20 X K-30-1)	13.73	10.00	29.00	--
Poornima	Hybrid (BLA 139-1 X K-30-1)	14.08	7.8	31.00	--

Note: *Figure in paranthesis indicates the age of the tree at which the yield was recorded. OP-Open Pollinated, T. No. - Indicates the Tree number.*

Drip irrigation @ 80 L of water per tree once in 4 days is also found to be encouraging.

Training and pruning: All the side shoots must be removed up to a height of atleast 2 m from the ground to cause the branches to form and spread out from the upper section of the trunk. Periodical pruning of dead wood and criss cross branches during the month of July is recommended to minimise the losses through diseases such as dieback and to increase the yield. As cashew bears its panicle terminally on the current season shoots, immediate after harvest pruning is recommended to encourage more shoots which will produce new flushes bearing panicles in December-January. This is highly essential under HDP and UHDP systems.

Manuring: Though a regular manurial schedule is not followed by most of the growers, experiments show that application of organic and inorganic manures are essential for higher yield in cashew. The recommended manurial schedule is as follows:

Age of plantation	Manures per tree			
	FYM or Compost (Kg)	N	P	K
		(g)		
One year old	10	50	25	25
Two years old	20	100	50	50
Three years old	20	150	75	75
Four years old	30	150	75	75
Five years and above	50	500	125	125

In places receiving both South West and North East rainfall, the above quantity can be split into two equal doses and applied and in areas receiving only North-East monsoon, the entire quantity is applied during November. Fertilisers may be applied within a radial distance of 2-3 m leaving half a meter from the trunk or in narrow trenches of 15 cm deep and covered with soil. The foliar spray consisting of NPK 19:19:19 @ 1 % at new flush stage (August), Mono-Ammonium phosphate @ 1 % + boron 0.1 % during flowering (December) followed by third spray of TNAU Panchagavya 3 % during fruit set stage (January-February)especially in pruned field is recommended to increase the yield substantially.

Cropping: The cashew tree commences fruiting in the third or fourth year, attains the full bearing age by the tenth year and lives for 40 to 50 years. Flowering commences in November and extends up to February. The peak months of harvest are March-April and the remaining crop comes to harvest in February and May.

The nuts collected should be dried immediately under sun by spreading in a thin layer. If the surface is of cement concrete, drying for two full days is sufficient. If the surface is of mud, drying for 3 to 4 days is necessary. While drying, the nuts should be raked quite often. Nuts should not be dried for more than four days since they become brittle and break while processing, causing damage to the kernels.

Yield

The yield depends upon many factors. Individual trees which yield more than 6 kg after 15 years are considered as good yielders.

Cashew Processing

Processing consists of roasting, shelling, extracting the oil, and peeling, grading and packing.

(a) **Roasting:** Roasting makes the shells brittle, besides making the extraction of kernels easier. A slight under roasting or over roasting adversely affects the quality and recovery of kernels. In the open pan roasting method, one kilogram of nuts is kept in shallow iron pans or earthen pots and is heated over an open fire. The nuts are rapidly turned to prevent charring. During the process of roasting, large quantities of shell liquid and smoke would come off and cause an irritation and injurious effect on the skin of the personnel engaged in the operation. The roasted nuts are then removed from the pan and thrown on the floor. They are quickly covered with earth which would absorb shell oil adhering to the roasting nuts and also cool them. The nuts are then subjected to subsequent operations. The cashew nuts are also roasted by drying under sun for two to three days when they lose much of the moisture contents and become brittle enough for shelling. The other improved methods of roasting cashew nuts are

(i) **Continuous roasting process:** The principle adopted in this system is the same as in the case of open pan roasting method. This plant consists of a single walled or double walled rotating metallic drum. In the case of double walled drum, the smoke or gases produced during roasting escape through the interspace between the two walls of the drumand are condensed to shell oil by a cooling system but in the case of single walled drum, the gases that escape from the nuts during the process of roasting are allowed to escape through a chimney provided at the lower end and there is no provision for collecting shell liquid.

(ii) **Oil bath process:** In this method, the nuts are held in wire trays and are passed through a bath of cashew shell oil maintained at a tempera ture of 200 to 202°C for a period of three minutes whereby the shell oil is recovered from the shells to the maximum possible extent. This process ensures uniform roasting of nuts and eliminates charring of kernels.

(b) **Shelling:** After roasting, shelling is done by labour. Each nut is placed edgewise and cracked open with a light wooden mallet and the kernel extracted with or without the help of a wire prong. Care has to be taken that the inner kernel is intact and is not broken into bits.

(c) **Peeling:** Removal of a thin outer brown skin is done by hand with the help of a safety pin or small hand knife. Peeling is made easier when the kernels are subjected to a heat treatment for about four hours in a drying chamber.

After peeling, the kernels are spread out indoors on cement flooring so that they may absorb some moisture and become less brittle. This prevents the tendency to break easily during grading.

(d) **Grading:** Grading is done based on "counts" or number of kernels per pound. The kernels which have no split are separated as 'wholes'. These are again separated into six grades as 210, 240, 280, 320, 400 and 430 whole nuts per pound. The graded kernels should be fully developed, ivory white in colour and free from insect damage and black or brown spots. The broken and split kernels are then separated and classified as standard and scorched pieces, splits, butts, small pieces and each grade is separately packed.

(e) **Packing:** Packing is done in this. In this method, the air inside the tin is exhausted and they are recharged with CO_2 before they are sealed air-tight.

Top Working

As most of the existing cashew plantations are of seedling progenies, the yield level is very low and highly erratic. Hence, top working with improved clones are suggested now. Trees of 20 to 25 years old are beheaded at a height of 0.5 m from the ground during December-February. A paste, made using 50 g each of BHC 50 per cent wettable powder and copper oxychloride in a litre of water, should be applied all over the stump to check any infection by invading pathogens and borer insects. Profuse sprouting normally results in but only 10 to 15 healthy shoots and properly spaced on the stumps are alone retained. These shoots are grafted ᵗt softwood stage (cleft grafting) when they are about 40 to 50 days old. 7-8 successful grafts may be encouraged to grow and the sprouts should be periodically removed. Top worked trees grow vigorously due to the well established root system and they start yielding about 4 kg per tree from the second year of rejuvenation and the yield gradually increases to stabilise at 8 kg from the fourth year of top working.

Plant Protection

The following two pests are economically important in cashew plantations.

1. **Stem borer:** The grub bores into the trunk and the roots. Control measures involve collection and destruction of the affected shoots, swabbing the bark of exposed roots and shoots with carbaryl 50 WP @ 2 g/L twice a year before the onset of South West Monsoon (March – April) and after cessation of monsoon (November) painting of coal tar + kerosene mixture (1:2), root feeding with monocrotophos 36 SL @ 10 ml + 10 ml of water kept in a polythene bag on one side of the tree and keep the same amount on the other side of the tree (Total 20 ml/tree) divided into two equal halves ,remove grubs from early stage infested trees and drench the damage portion with Chlorpyriphos 0.2 % @ 10 ml/L or Neem Oil 5 %.

2. **Tea mosquito bug:** Adults and nymphs suck the sap from the tender plant parts. Spray application of fipronil 5 SC @ 1.0 ml/L followed by carbaryl 50WP @ 2 g/L and monocrotophos @ 2 ml/L at vegetative flush stage, panicle initiation stage and nut formation stage respectively. Spray schedule involving three rounds of spray *viz.,* first spraying with Profenophos (0.05 %) at flushing stage, second spraying with Chlorpyriphos (0.05 %) at flowering and third spraying with Carbaryl (0.1 %) at fruit set stage is most effective.

3. **Die back or Pink disease:** Spraying any copper fungicides besides pruning the dead twigs are suggested.

CHAPTER

14

Coconut

COCONUT (*Cocos nucifera* L.)

Family: **Palmae**

The coconut palm, *Cocos nucifera L.,* is one of the most beautiful and useful palms in the world. It provides a variety of useful products like food, fuel and timber. Every part of the tree is being utilised for some purpose or other and hence, it is called **Kalpavriksha** meaning **tree of heaven** which provides all the necessities of life.

It is grown in India in about 1.78 million ha. with an annual production of about 12832.9 million nuts contributing 16.70 % in world production .It ranks third in world in area and production, first and second being Indonesia and Philippines in terms of production respectively but in terms of area it is *vice versa*. Among the different coconut growing states in India, Kerala, Tamil Nadu, Karnataka and Andhra Pradesh account for nearly 85 per cent in area and production.

Name of the State	Share in area (%)	Share in production of nuts (%)
Kerala	37.06	28.37
Tamil Nadu	20.63	26.35
Karnataka	24.85	26.91

Andhra Pradesh	6.96	9.06
Other (Orissa, Maharashtra, Assam, W. Bengal, Goa, Daman, Andaman, Pondicherry, Tirupura etc.)	9.98	9.28

Botany

Origin of coconut is believed to be somewhere in South East Asia. Coconut, botanically *Cocos nucifera* has only one species under the genus *Cocos*. It is a tall, stately unbranched palm growing to a height of 12 to 24 m. The stem is marked by rings of leaf scars which are often not prominent at the base.

The palm has an adventitious root system, having numerous thick roots from the base of the stem almost throughout its life. The roots are localised generally at the lower most region of the stem which has been termed the bole.

Leaves are large, long, pinnatisect, borne on the crown. The palm is monoecious with relatively few female flowers. Male flowers are numerous small with 6 stamens and in female flowers, the ovary is tricarpic, usually one ovuled.

Fruit is a large, one seeded drupe. The outer layers of the pericarp are thick and fibrous. The inner layer (endocarp of shell) is very hard, horny or stony and the thin testa cohering to the endocarp is lined with white albuminous endosperm (meat), enclosing a large cavitiy, partially filled with sweet fluid.

The inflorescence develops within a strong, tough pointed double sheath called spathe which after full development splits along its underside from top to bottom and releases the inflorescence. This usually occurs from 75 to 90 days after the first appearance of its tip in the leaf axil. The primordia of the inflorescence begin to form in the leaf axil about 32 months before the opening of the spathe. In bearing coconut palm every leaf axil can produce a spadix and under normal conditions it varies from 12-15 per annum. However, this number may be reduced due to adverse weather condition.

In India, the female flower production is high during the period from March to May and low from September to January. In general, the number of female flowers per inflorescence varies from 10-50. Female flowers normally become receptive 19 to 20 days after the opening of the spathe. Genetically, the dwarf palms are autogamous while tall types are allogamous. Both winds and insects are considered to be the main pollinating agents. A large number of buttons (Female flowers) fail to develop into nuts due to lack of pollination and fertilization, defects in the flowers, physiological disorders, genetic nature of the variety, pest and disease and unfavourable environment etc. Generally not more than 25 to 40 per cent of the female flowers reach maturity under normal conditions.

Climate and Soil

The coconut palm is found to grow under varying climatic and soil conditions. It is essentially a tropical plant, growing mostly between 20° N and 20° S latitudes. Near the equator, productive coconut plantations can be established up to an elevation of about 1000 metres from sea level while the farther one goes from the equator, the more is the palm confined to low lands. The palms tolerate wide range in intensity and distribution of rainfall. .However, a rainfall of about 200 cm per year and well distributed throughout the year is the best for proper growth and maximum yield. In areas of inadequate rainfall with uneven distribution, irrigation is required. Soil moisture deficit during summer hampers nut production to greater extent. Palm requires plenty of sunlight and does not grow well under shade. Minimum of 4 hrs of sunshine per day is essential and places with relative humidity 80-90 % are ideal, however, it should not go below 50 %.

Coconut is adaptable to a wide range of soil conditions, from light sandy soils to heaviest clays with a pH ranging from 5.2 to 8.5. Best soils are deep, friable, loamy soils. In heavier soils, it requires good drainage.

Cultivars and Hybrids

Coconut palms are broadly classified into two groups, the tall and dwarf. The tall cultivars are the common type that occurs throughout the world. The different cultivars of the Talls are known by the place where they are largely cultivated. The tall cultivars largely grown in India are the West Coast Tall and East Coast Tall. The dwarf varieties are shorter in stature and life span as compared to Talls. They start bearing earlier compared to Talls. The size of the nuts and the quality of copra are inferior to Talls. The Dwarf cultivars occur with three nut colours viz. green, yellow and orange. The dwarf cultivars are generally grown for tender nuts and also for hybrid production. The common dwarfs available in India are Chawghat Orange Dwarf, Chawghat Green Dwarf, Malayan Green Dwarf, Malayan Yellow Dwarf, Malayan Orange Dwarf, Gangabondam etc.

The hybrids between Tall and Dwarf forms (TxD) or *vice versa* (DxT) show hybrid vigour for growth, earliness and yield. Hence, hybrids have been released recently for cultivation. The important cultivars and hybrids recommended now for cultivation are given below:

Cultivar/hybrid	Time taken for flowering (months)	Nut yield per palm/ year (number)	Copra content (g)	Oil content (%)
Tall				
West Coast Tall (**WCT**)	60.8	80	176	68
East Coast Tall (**ECT**)	72.9	73	125	64
Kalpa Dhenu	67	22,794 **	–	–
Kalpa Prathibha	–	23,275**	–	–
Kalpa Mitra	–	180	–	–
Chandra Kalpa	–	100	–	–
Kalpa Tharu	–	116	176	–
Kalpa Tharitha	–	118	25.5***	–
Dwarfs				
Kalpa Raksha	54	87	16.38 ***	10.65 ***
Chowghat Orange Dwarf (COD)	3-4 *	36	–	–
Kalpa Sree	2.5–3 *	90	96.3	–
Kalpa Jothi	–	114	16 ***	–
Kalpa Surya	–	123	23 ***	–
Hybrids				
Kera Sankara (WCT × COD)	48.6	108	187	68
Chandra Sankara (COD × WCT)	60.0	116	215	68
Chandra Laksha (L.O × COD)	48.0	109	195	69
Laksha Ganga (L.O × G.B)	60.0	108	195	70
VHC-1 (ECT X MDG)	48.0	98	135	70
VHC-2 (ECTx MDY)	48.0	107	152	69
VHC-3 (**ECT X MOD**)	–	156	161.5	64.5
Kera Ganga (WCT × G.B)	60.0	100	201	69

Anantha Ganga (A.O × G.B)	60.0	95	216	68
Godhavari Ganga (ECT × G.B)	37.5	140	150	68
Kalpa Samrudhi (MYD x WCT)	–	117	4.38[@]	3.04[@]
Kera Sree (WCT X MYD)	–	112	216	66
Kera Sowbagya (WCT X SSAT)	–	130	195	65
Kalpa Sankara (CGD x WCT) (recommended for cultivation in root wilt disease prevalent tract of Kerala)	–	85	2.5[@]	1.6[@]

N.B. : COD - Chowghat Orange Dwarf, CGD- Chowghat green Dwarf, A.O - Andaman Ordinary, L.O- Laccadive ordinary, G.B - Ganga Bondam, MDG- Malaysian Dwarf Green, MDY- Malaysian Dwarf Yellow, MOD- Malaysian orange Yellow

* years, ** nuts/ha/year, *** kg/palm/year, [@] tones/ha

Further, many varieties have been developed through exotic and indigenous germplasm selection in India as indicated below:

Cultivar	Bearing age (Years)	Annual nut yield	Annual copra yield	
		Nuts/palm	g/nut	Kg/palm
Exotic types				
Fiji Tall	6	106	199	21.1
Philippines Ordinary (Kera Chandra)	5	110	198	21.8
Strait Settlement Green	6	108	186	20.0
Seychelles (Kera Sagara)	8	99	203	20.2
Indigenous types				
Lakshadweep Ordinary (Chandra Kalpa)	6	100	169	17.6

Andaman Ordinary (VPM3)	6	94	152	16.0
Benaulim (Pratap)	7	150	172	22.8
Chowghat Orange Dwarf (COD)	7	80	–	13.8
Assam Tall (Kamrupa)	4	63	162	–
Arasampatti Tall (ALR-1)	5	101	131	–
West Coast Tall (WCT)	5	176	-	–

Production of Planting Material

Coconut being a perennial, cross pollinated crop with its economic life span lasting for 60 years or more, it is important that only genetically superior seedlings are planted in the field. This involves multiple steps, starting from selection of mother palms. Good mother palm can be identified by the following traits:-

1. Crown should be spherical or semi-spherical, drooping or erect crown should be avoided.
2. Palm should have 30-40 fully opened leaves and 12 to 15 bunches with a high setting of female flowers.
3. Nuts should be medium in size and nearly round or spherical in shape.
4. Palm should be in the age group of 25-50 years.
5. Palms growing close to house, cattle sheds, compost heaps etc. should be avoided.
6. Mother palm selected should produce more than 120 nuts/year
7. Dehusked nut should weigh more than 600 g with copra more than 150 g
8. Mother palms selected should be free from pest and diseases.

Nuts from such selected mother palms when attain full maturity (11 to 12 months) are harvested along with the bunches. These nuts are arranged with their stalk ends up on the floor of a shed over a layer of about 7.5 cm dry sand and completely covered with it till the planting time.

The nursery site is selected in a sandy area otherwise; soil to a depth of 0.3 to 0.5 cm is removed and filled with sand. The raised seed beds are prepared to convenient length but the width is so adjusted to take 4 or 5 rows of nuts only.

Nuts can be planted both vertically and horizontally, but the latter one is preferred since the seedlings develop from such plantings are robust, suffer lesser damage in transist, besides exhibit higher germination and vigorous growth. The depth of planting should be so adjusted that the husk just appears above the surface.

The nuts generally commence germination in 11 to 12 weeks after planting. The seedling for planting should be atleast 9-12 months old and are selected by looking for the following traits:

(a) Early and vigorous seedlings.

(b) Seedlings having maximum girth at the base.

(c) Seedlings with early splitting of leaves.

Recently, seedling production is also taken up through polybag system. Polybags of 60 × 40 cm size - 500 gauge thicknesses with 8-10 holes are filled with a mixture of top soil, sand and manure in 2:1:1 ratio. Just sprouted seedlings are set in this bags; the main advantage are

• Least damage during planting

• Quick growth after planting

• Lesser mortality rate

Preparation of land and planting: Preparation of land for planting coconut depends to a large extent on soil type and environment at factors. The depth of pits will depend upon the type of soil. Normally a pit size of 1.0 × 1.0 × 1.0 m is dug and filled up to 50 cm depth with sand and powdered cow dung. However, when the water table is high, planting at the surface or even on mounds may be necessary but digging pits and filling has to be done.

Spacing of palms requires careful consideration. A spacing 7.5 to 9.0 m may be adopted depending on the crown size. This will accommodate 177 to 124 palms per ha under the square system of planting and an additional of 20 to 25 palms can be planted if the triangular system is adopted. A hedge system can be also adopted with a spacing of 5.0 to 5.5 m along the rows and 9 to 10 m between the rows. This system provides ample opportunity to grow a number of perennial and annual crops in the interspaces.

In well drained soils, seedlings can be transplanted with the beginning of south-west monsoon. If irrigation facilities are available, it is advisable to take up planting at least a month before the monsoon sets in so that the seedlings get well established before the onset of heavy rains.

Care of Young Palms

Young palms require good care in the early years of growth. The transplanted seedlings should be shaded and irrigated properly especially during the summer months. Provision of proper drainage is also equally important in areas subject to water logging.

The pits should be cleared of weeds periodically. Soil washed down by the rains and covering the collar of the seedlings should also be removed. The pits should be gradually filled up as the seedlings grow.

Manuring: As coconut yields throughout the year, it takes heavy amount of nutrients from soil especially N, K and Cl. Therefore, regular manuring from the first year of planting is essential to ensure good vegetative growth, early flowering and bearing and high yields. Organic enrichment in the basins of the palm is highly recommended. Green manure crop like (leguminous) *Pueraria, Calopagonium, Mimosa,* Cowpea and Sunhemp are raised thickly and just before ploughing they are incorporated. Green leaf manure from gliricidia can be also incorporated @ 10 t/ha to improve the fertility of the soil. Recently CPCRI has developed the techniques of making compost from coconut fronds using earth worms. Dried fronds for 1-2 months are made into small pieces and mixed with 100 kg of cattle manure to encourage the worms. A moisture content of 30-40 % should be ensured; within a period of 90 days the coconut leaves get converted into compost. This compost contains about N (1.2-1.8 %), P (0.1 -0.2 %) and K (0.2-0.4 %) besides some micro nutrients.

The fertilizer requirement of different coconut growing states is given below:

State	Doses (g)/palm				
	FYM (kg)	N	P_2O_5	K_2O	MgO
Kerala-General	50	500	320	1200	–
Kerala-Root wilt prone palms	50	500	500	1000	500
Karnataka	50	680	450	900	–
Maharashtra	50	750	225	900	–
Tamil Nadu	50	560	320	1200	–

The first application of fertilizers should be done three months after planting (nearly 1/10th above dose) while during the second year, one-third of the dosages recommended for adult palm may be applied in two split doses and during the third year, 2/3 of the dosages recommended for adult palms may be applied and from the fourth year onwards, the full dose recommended may be applied.

After the receipt of summer showers, one-third of the recommended dose of fertilizers for a year may be spread around the palms within a radius of 1.8m and forked in. The remaining two-third of the recommended dose of fertilizers may be spread over 50 kg of the green leaf or compost per palm in circular basins of 1.8m radius and 25 cm depth during September/October and the basins covered.

Irrigation and soil moisture conservation: The coconut palm responds to summer irrigation. Production of female flowers and setting percentage increases considerably due to irrigation. Since spadix initiation to ripening of nuts takes nearly 42 months, the full benefit of irrigation can be felt only after 3 years. Under West Coast conditions, 2 cm irrigation once in 4-5 days during December-May is beneficial in sandy loam soils. In areas where water is scare, drip irrigation system can be adopted. This requires only 30 to 75 litres of water per day per palm as against 200 litres per palm in conventional method.

Burying the coconut husk or coir dust is one of the most effective ways of conserving soil moisture. These husks or coir dusts can act as sponge and absorb and retain moisture about six and ten times respectively to their own weight and slowly release to the coconut trees during dry periods. As the husk or dust breaks down slowly, their effect will last for 4 to 6 years and 8 to 10 years respectively. On decomposition they also add potash to the soil. These husks or dusts can be added in pits or trenches taken in between the trees but in all the cases the depth should be 0.6 m and 1.8 m away from the bole. Husks or dusts can be added in alternate layers with soil. Coir dust compost can be also added which also supply NPK besides enhancing the soil.

Intercultivation: Regular intercultivation is very essential to step up and maintain the production at a high level. Tillage operations like digging the garden with mammutty (spade), ploughing, forming small mounds before the end of monsoon and making shallow basins with a radius of about 2 m at the beginning of monsoon and filling up at the close of monsoon are beneficial to the trees. Method of intercultivation will depend upon local conditions, availability of labour, size of holding, soil type, topography and distribution of rainfall.

Cover cropping is recommended where inter and mixed cropping is followed to prevent soil erosion in coconut gardens. Leguminous crops such as *Mimosa invisa, Stylosanthes gracilis* and *Calopogonium mucunoides* are generally recommended. Green manure crops like sun-hemp *(Crotolaria juncea)* and *kolinji (Tephrosia purpurea)* are also raised and ploughed in during August - September. These crops can be sown in April - May when pre monsoon showers are received.

Inter and mixed cropping: In pure coconut garden when palms are spaced at 7.5 × 7.5 m as much as 75 percent of the available area is not effectively utilized. Besides, a pure coconut grove utilizes only half of the available light. Hence, a variety of intercrops like pineapple, banana, and elephant foot yam, groundnut, chillies, sweet potato and tapioca can be raised in coconut gardens after the palms attain a height of 5 to 6 meters. In older plantations, cocoa, pepper, cinnamon, clove and nutmeg can be grown as mixed crops. In places where rainfall is not well-distributed, irrigation may be necessary during summer months. However, these crops are to be adequately and separately manured in addition to the manures applied to the coconut palms.

Mixed farming by raising fodder grasses such as hybrid napier or guinea grass along with leguminous fodder crops such as *Stylosanthes gracilis* in coconut gardens can support four to five dairy animals. The cattle manure generated from the system when applied to coconut garden improves the soil fertility considerably.

Harvesting

Fully matured nuts which can be recognized by shaking the nuts should be alone harvested to get the maximum yield of copra and oil. Frequency to harvest varies from place to place. In any parts of Kerala, harvesting is done at 45 days interval during summer months and at 60 days interval during the rainy season. Hence, 35 percent of the total nut is obtained during the hot months i.e., from March-May and least crop is obtained during rainy months. Harvesting is done by climbing the tree.

Tree Climber

From time immemorial only skilled labours were engaged in climbing coconut palms for harvesting. This operation is tedious, risky and requires lot of skill. Due to various socio- economic factors the number of such traditional skilled palm climbers is steadily declining. In this context, mechanical climbing device has been introduced. Coconut Development Board through an innovative scheme 'Friends of coconut tree' provides adequate training to youth and women so that they are well trained to use the climber for harvesting.

Yield

It is influenced by the agro-climatic conditions, cultural practices such as manuring, irrigation, cropping system and control of pests and diseases etc. On an average, a tall variety yielding 60-80 nuts per palm per year is considered ideal.

Coconut Products

Products of coconut can be broadly classified into three categories *viz.*, food products, commercial products, coconut shells and miscellaneous products.

A. Food products

(i) Wet meat or Kernel (endosperm) is used for culinary purposes in different forms. It contains proteins (4.5 %), fat (41.5 %), carbohydrates (13 %), fibre (3.6 %) and minerals (1 %). With maturity fat increases and protein and mineral decrease.

(ii) Coconut milk and related products - the milk obtained by squeezing grated kernel is used in preparation of various food products.

(iii) Coconut milk powder - the coconut milk formulation is filtered, homogenized and pasteurized and the product is spray dried at 100-150^{0}C to produce coconut milk powder which easily gets dissolved in water to form milky white liquid with flavor and texture of coconut milk.

(iv) Desiccated coconut- the dried, disintegrated , desiccated at 77-82°C, coconut meat of fully matured nut stored for a month before dehusking which has a moisture content of 2- 2.5 %.

(v) Coconut flour - is obtained by partial removal of fat by hydraulic pressing from the coconut gratings which are dried and powdered.

(vi) Coconut water - liquid endosperm of tender coconut (5-7 months) makes a refreshing drink and is rich in Vitamin C, B and minerals especially potassium.

(vii) Toddy is the sweet juice containing sucrose obtained by tapping the unopened spadix. The flow of sap continues for one month and 15-18 litres can be extracted in a month. Fresh toddy when allowed to ferment turns the sugar into alcohol (5-8 %).

B. Commercial products

(i) Milling copra - (6 % moisture content) obtained from fully mature nuts (45-55 % moisture content) is used for production of oil. Sun drying is the cheapest method to produce this and it takes 5-7 days. Solar tunnel dryers are now used which is very efficient and the product is less infested by fungi and bacteria.

(ii) Coconut oil - extracted by crushing copra in rotary mills or expellers and the oil content ranges from 65-75 %.

(iii) Virgin coconut oil - the naturally processed product of fresh coconut meat either directly or through extraction of coconut milk. It is colourless and has a mild to intense coconut scent and is rich in lauric acid (47-53 %) and Vitamin E (Tocopherol) at 5 mg/kg.

C. Coir and Coconut fibre

The coconut husk contributes 30 % of coconut fibre. There are two types of fibres *viz.,* white and brown fibre.

White Fibre- extracted from green husks after retting in natural water for 6-10 months.

Brown fibre- extracted from dry husks by mechanized defibring process.

The waste product of coir industry is called as 'cocopeat' which is used as mulch material and can be converted into compost also.

Plant Protection

The major pests and diseases affecting the coconut palm are furnished below:-

Pests/diseases	Symptoms/damage	Control measures
Rhinoceros beetle (*Orycetes rhinoceros*)	Adult beetle bores into the unopened fronds and spathes, the affected fronds when fully opened show characteristic geometric cut.	Proper disposal of decaying organic debris extracting the beetle with a hook mechanically, filling the inner most three or four leaf axils of palm with a mixture of 5% malathion dust and sand (1:1), three times a year (April, Sept. and December). Spray *Metarhizium anecopliae* culture in the compost yard, keep the sex trap @ one/ha to destroy these beetles
Leaf eating caterpillar	Caterpillars live on the undersurface of	Spraying of infected palms on the lower surface of the leaves with

(Opisina arenosella)	leaflets inside silken galleries and feed on chlorophyll.	Dichlorvos (0.02 %) or 5 % azadirachtin based neem product @ 7.5 ml along with 75 ml of water and give root feeding. Biological control through the release of parasites like *Goniozus nephantidis*, *Bracon brevicornis*,*Tetrastichus*, *Israeli* and *Trichospilus pupivora*
Red Palm weevil (*Rhynchophorus ferrugineus*)	Presence of hole, Oozing out of a viscous brown fluid and extrusion of chewed up fibres through the sides, longitudinal splitting of leaf bases and wilting of inner leaves of the palm of 5-20 years age group.	Injection of Pyrethrin piperonyl butomide (Pyrocon-E) 10ml in one litre of water per palm into the trunk through a hole above the infested portion, provision of traps by keeping fresh toddy, fermented with yeast and acetic acid etc., keep pheromone trap (Ferrolure) at 2 no. s/acre
Rats	Damage tender nuts and cause severe crop loss.	Provide mechanical barrier (G.I. sheets of bands 40 cm wide fixed around the trunk at 2m high), poisonous baiting with anti blood coagulant like Warfarin/fumarin compounds.
Diseases		
Bud rot (*Phytophthora palmivora*)	Yellowing of one or two young leaves surrounding the spindle and they wither and droop drown often emitting a foul odour.	Application of Bordeaux paste in the cut portion of the infected tissues at early stage of infection itself. Prophylatic spraying of 1 % Bordeaux mixture to all healthy palms.
Root (wilt) diseases.	Prevalent throughout Kerala, adjoining districts in Tamil Nadu, most visual symptoms is abnormal bending of ribbing of the leaflet, a general yellowing, marginal necrosis of leaflets, yield is reduced considerably.	No curative control measure is known. Removal of the affected palms and replanting with hybrids. Application of recommended N, P, K, MgO and organic manures, Growing of green manure crop at basins and irrigation during summer.

Stem bleeding (*Thielaviopsis paradoxa*)	Exudation of a reddish brown liquid through cracks on the trunk later the bleeding points start decaying.	Remove the affected tissues using a chisel and dress the wounds with hot coal tar and or Bordeaux paste.
Thanjavur wilt (*Gonoderma lucidum*)	Decay of root system, flaccidity of spindle fibres, browning of outer leaves, arrested fruit set, appearance of bleeding patches on the stem, affected palms die within 2-3 years.	Apply 5 kg of neem cake per year, addition of organic matter, providing irrigation could check the spread of disease. Aureofungin – sol 2 g + one g Copper sulphate or 2 ml of Tridemorph dissolved in 100 ml water may be applied as root feeding. The active absorbing root of pencil thickness be selected and a slanting cut is made. The solution is taken in a polythene bag or bottle and the cut end of the root is dipped in the solution. Forty litres of 1% Bordeaux mixture should be applied as soil drench around the trunks in a radius of 1.5 metre

CHAPTER

15

Arecanut

ARECANUT (*Areca catechu* L.)

Family: **Palmae**

Arecanut palm is cultivated primarily for its kernel obtained from the fruit which is chewed in its tender, ripe or processed form. Though it is native of Malayan Archipelago, Philippines and other East Indian Islands, commercial cultivation is confined only in India, Bangladesh and Sri Lanka. India has about 3.96 lakh hectares under this crop with an annual production of 5.59 lakh tonnes accounting world's largest producer in terms of area (62 %) and in production (60 %). Kerala, Karnataka and Assam account for more than 90 per cent of the total area and production in our country.

Arecanut production in India has now almost reached a level of self-sufficiency. Uses for arecanut other than chewing are negligible. Its export prospects are also very much limited. Therefore, the present policy is not to expand the area under arecanut, but to adopt intensive cultivation and take up replanting of the aged and unproductive gardens. Inter and mixed cropping in arecanut gardens is advocated to augment the income from the existing arecanut garden.

Botany

It is a monoecious palm and its inflorescence is a spadix produced in the leaf axil and is completely enclosed in a sealed boat shaped spathe. The spadix is having a main rachis divided subsequently into secondary and tertiary rachies. Female flowers are confined to tertiary and distal end of the secondary rachis, while male flowers are produced on filiform branches arising below and beyond the female flowers. Both female and male flowers are sessile, with two whorls of perianth. The fruit is a monolocular, one seeded berry and it consists of a fibrous outer husk, enclosing a single seed. It is a cross pollinated crop and fruit set normally varies form 12.0 to 40.0 percent and the time taken from full bloom to maturity of the fruit ranges from 35 to 47 weeks.

Climate and Soil

The arecanut palm is capable of growing under a variety of climatic and soil conditions. It grows well from almost sea level up to an altitude of 1000 m in areas receiving abundant and well distributed rainfall (750 mm to even up to 4500 mm) or under irrigated conditions. Although it can tolerate extreme temperature, the optimum range will be around 14-36 ° C. It is grown in soils such as laterite, red loam and alluvial soils. The soil should be deep and well drained with a pH ranging from 5-7.

Varieties

There are few local varieties known by the name of the place where they are grown and are furnished below:-

Name of the local variety	Place where grown
South Kanara	Dakshina Kannada district and Kassargod district of Kerala
Thirthahalli	Malnad area of Karnataka
Sreevardhan	Coastal Maharashtra
Mettupalayam	Coimbatore District
Mohitnagar	West Bengal
Kahikuchi	Assam
Hirehalli Dwarf	Karnataka

Central Plantation Crops Research Institute, Regional Research Station, Vittal has released five improved cultivars and two hybrids and are briefly described below:

Name of the cultivar/Hybrid	Special attributes
Mangala	An introduction from China (VTL-3) early bearing, higher fruit set, higher yield (10 kg ripe nuts/palm/year), semi tall variety.
Sumangala	A selection from Indonesia (VTL-11), yield 17.5 kg of nuts/palm at the age of 10 years.
Sreemangala	A selection from Singapore (VTL-17), yields 16.5 kg/palm at the 10th year.
Swarnamangala	A selection from high yielding variety VTL-12 (Saigon), semi tall variety, regular bearer, high recovery of chali (26.40 %), yields 3.70 kg chali/palm/year.
Mohit Nagar	A selection from West Bengal type, yields 3.67 kg chali/palm/year.
VTLAH1 (Vittal Arecanut hybrid 1) Hirehalli Dwarf X Sumangala	Dwarf nature with sturdy stem with superimposed nodes, reduced canopy size; well spread leaves, partial drooping crown, medium size oval to round and yellow orange colored nuts. High recovery of chali (26.45 %). Average chali yield 2.55 kg/palm/year.
VTLAH2 (Vittal Arecanut hybrid 2) Hirehalli Dwarf X Mohitnagar	Dwarf, reduced canopy, early yield stabilization, oval, medium, orange nuts. Average yield (Chali/palm): 2.64 kg
Other New varieties	
Samruthi	This variety is recommended for Andaman and Nicobar islands, high yield potential of 18.89 kg ripe nuts/palm/year with chali yield 4.34 kg/palm. Well spaced bunches around and bold nuts.
SAS1 (Sirsi Arecanut Selection 1)	This variety is recommended for hill zone of Karnataka, tall with compact canopy, regular bearing variety. Suitable for both tender and ripe nut processing. High curing percentage and yields 4.60 kg chali/palm/year.

Raising of Planting Materials

Collection of seed nuts should be confined to high yielding palms which commence to bear early as well as those which give more than 50 per cent of fruit set. From these selected mother palms, fully ripe nuts are alone collected. All undersized and malformed nuts must be rejected. Heavier seed nuts (above 35 g) within a bunch are alone selected, as they give higher percentage of germination and produce seedlings of better vigour than lighter ones.

The selected seed nuts are sown immediately after harvest, 5 cm apart in sand beds under partial shade with their stalk ends pointing upwards. Sand is spread over the nuts just to cover them. The beds may be watered daily. Germination commences in about 40 days after sowing and the sprouts can be transplanted to the second nursery when they are about three months old. At this stage the sprouts might have produced two to three leaves.

The secondary nursery beds of 150 cm width and of convenient length are prepared for transplanting the sprouts. The sprouts are transplanted at a spacing of 30 cm × 30 cm with the onset of monsoon. Partial shade to the seedlings can also be provided during summer by pandal or growing banana. Care should be taken to drain the nursery beds during the monsoon and to irrigate them during the dry months. Weeding and mulching should be done periodically. Seed nuts can also be sown in polythene bags (25×15 cm size, 150 gauge) after filling the bags with potting mixture containing 7 parts of loam or top soil, 3 parts of dried and powdered farm yard manure and 2 parts of sand.

The seedlings will be ready for transplanting to the main field when they are 12 to 18 months old. Seedlings having 5 or more number of leaves should be selected. The height of seedlings at the time of planting has a negative correlation with the subsequent yield of the plant. Hence shorter seedlings with maximum number of leaves are removed with a ball of earth for transplanting. If the seedlings are raised in polythene bags, these can be straightway transported to any distance without much damage.

Planting

The planting is done during May-June with the onset of monsoon. Arecanut palms need adequate protection from exposure to the south western sun as they are susceptible to sun-scorch. Proper alignment of the palms in the plantation will minimize sun scorching of the stem. In the square system of planting at a spacing of 2.7 m × 2.7 m, the north south line should be deflected at an angle of 35° towards west. Dwarf cultivars and hybrids may be planted at 2.2 m × 2.2 m. The outermost row of plants on the southern and south-western sides can be protected by covering the exposed stem with areca leaves or leaf sheaths or by growing tall and quick growing shade trees.

Pits of 90 × 90 × 90 cm are dug and the pits are filled with a mixture of top soil, powdered cowdung and sand to a height of 50 to 60 cm from the bottom. The seedlings are planted in the centre of the pit, covered with soil to the collar level and pressed around. A shade crop of banana can be raised to give protection to the seedlings from sunscorch.

Manuring

Adequate supply of plant nutrients in the soil throughout the life of the crop is essential to get high yield. Hence, an annual application of 100 g N, 40 g P_2O_5 and 140 g K_2O in the form of fertilizers and 12 kg each of green leaf and compost or cattle manure per bearing palm is recommended. Under rainfed conditions, half the quantity of fertilizers may be applied in April-May and the remaining quantity in September-October. Under irrigated condition, the first dose of fertilizer is applied only in February. Green leaf and compost can be applied in single dose in September-October. Irrespective of the age of the plants, full dose of green leaf and compost or cattle manure may be applied from the first year of planting itself while one-third of the recommended quantity of fertilizers in the first year, two-third in the second year and the full dose from the third year onwards. The first dose of fertilizers may be broadcast around the base of each plant after weeding and mixed with the soil by light forking, while the second dose is done in basins around the palm dug to a depth of 15 to 20 cm and at 0.75 to 1 m radius. In acidic soils, required quantity of the lime may be applied during the dry months and forked in. Fertigation is recently practiced where the fertilizers are applied during summer irrigation at 10-20 days interval by splitting the entire quantity of fertilizers in 9 or 18 splits.

Irrigation and drainage: The palms should be irrigated once in four to seven days depending on the soil type and climatic factors. In Kerala, arecanut gardens are irrigated during dry months once in seven or eight days during November-December, once in six days during January-February and once in 3-5 days during March, April and May. Adequate drainage should be provided during monsoon since the palms are unable to withstand water logging. Drainage channels should be 25 to 30 cm deeper than the bottom of the pits to drain excess water from the plot.

Other operations: A light digging may be required when the monsoon ends to break up any crust formed at the soil surface and also to uproot weeds. Weeding should be done periodically to keep the garden clean.

Cover cropping: Cover crops, such as *Mimosa invisa, Stylosanthes gracilis* and *Calapogonium muconoides* have been found to be suitable for arecanut gardens. The cover crops may be sown in the month of April-May and the green matter may be cut and applied to arecanut palms at the time of second dose of fertilizer application.

The crops which can be grown successfully in arecanut gardens without loss of arecanut yields are banana, cocoa, pepper, pineapple, betel vine, elephant foot yam, tapioca, dioscorea, sweet potato, arrow root, ginger, turmeric and guinea grass. Nutmeg and clove can be also grown in between four palms on alternate rows. In all the cases, manuring to intercrops should be also done separately.

Harvesting and Processing

The stage of harvesting depends on the type of produce to be prepared for the market. The most popular trade type of arecanut is the dried, whole nut known as chali or kottapak. Fully ripe, nine months old fruits having yellow to orange red colour is the best suited for the above purpose. Ripe fruits are dried in the sun for 35 to 40 days on dry leveled ground. For drying and dehusking, sometimes fruits are cut longitudinally into two halves and sundried for about 10 days, the kernals are scooped out and given a final drying.

Another form of processing is by making kalipak. The nuts of 6 to 7 months maturity with dark green colour are dehusked, cut into pieces and boiled with water of dilute extract from previous boiling, a kali coating is given and dried finally. *Kali* is the concentrated extract obtained from boiling 3 to 4 batches of Kalipak. Many varieties of scented suparis are now prepared by blending the dried, broken bits of arecanut with flavour mixtures and packed.

Dehusking of arecanut is traditionally done by skilled manual labour with the help of a tool which has a sickle shaped small pointed blade fixed on a plant. A simple device for dehusking arecanut, developed by CPCRI, Kasargod can also be used. The main advantage of this device is that any unskilled person can operate it. The outturn is about 60 kg husked nuts in case of dried nuts and 30 kg in case of green nuts if one person operates the device for a day of 8 hours.

In view of the declining use of arecanut as masticatory, alternate and better uses of arecanut and the other produces like spathe, husk are now thought of. One of the uses is to make areca plates from the spathe. It is eco-friendly, decomposes easily, light weight and easy to store. The demerits are that they are prone to fungal attack and can not be stored for long period. The areca spathes are collected and dried in sun for 3 days and are washed with power operated sprayer to remove all dirts and graded based on visual Such spathes are made into plates of different shapes (round, square and rectangular etc.).

Yield

More than 10 kg of ripe nuts per palm at the 10th year is considered as normal yield in any plantation.

Plant Protection

Pest/diseases	Symptoms	Control measures
Pests		
Mites (*Raoiella indica*) (*Oligonychus indicus*)	Adults and young ones suck the lower surfaces of the leaves, causing them to turn yellow and bronzed in appearance.	Spray the lower surface of leaves with fenazaquin 10 EC @ 1 ml/L
Spindle bug (*Calvalhoia arecae*)	Adults and young ones suck the sap from the tender spindle resulting in loss of vigour and consequent death.	Place 2 g of phorate granules taken in a perforated poly bags in the inner most leaf axils.
Inflorescence caterpillar (*Tirathaba mundella*)	Caterpillars feed on the flowers and clamp the inflorescence into a wet mass of frass with silky threads.	Infected spadices may be forced open and sprayed with malathion 0.05%.
Diseases		
Koleroga or mahali (*Phytophthora arecae*)	Water-soaked lesions appearing on the nut surface near the perianth spread over the other parts giving the nut a dark green colour. Infected nuts shed without perianth.	Spray 1 % Bordeaux mixture twice at 45 days interval during monsoon.
Bud rot (*Phytophthora palmivora*)	Affected spindle appear yellow, later changing to brown and finally the whole spindle rots.	Early removal of the infected tissue and treat the healthy tissue with 1 % Bordeaux paste. Drench the crown with 1 % Bordeaux mixture as a prophylactic measure.
Anabe roga (*Gonoderma luciderm*)	Small brown irregular patches appear on the stem and a brownish exudate oozes out from these patches.	Provide better, isolate the drainage affected palms by trenches, drench with 1 % Bordeaux mixture. Soil application of neem cake at 2 kg/palm/yr followed by root feeding with 125 ml of 1.5 % (15 ml/L of water) tridemorph @ 3 months interval

Yellow leaf disease (**Phytoplasma like organisms**) transmitted by the plant hopper (**Proutista**)	Leaves become yellow, smaller, stiff and pointed, crown gets reduced, palm remains stunted with few or no nuts.	Regular manuring, ensure proper drainage, grow cover crops, remove the affected palms.

Disorders

Areca palms exhibit following disorders:

1. **Sun scorching**

   Constant exposure of the stem to solar radiation causes this scorching. Golden yellow patches appear on the exposed portions of the stem, fissures develop later. Further, colonization by saprophytic organisms and insects cause decay of the stem and such palms break during heavy wind. Adoption of proper alignment in the north south direction while planting will greatly minimize the damage due to sun scorch. Growing banana as a shade tree in young gardens and trailing pepper vines on palms will reduce the radiation effect. Covering the trunk with areca leaf sheath and painting with lime are being practiced in few gardens. Palms with longitudinal fissures are to be reinforced with split areca stem.

2. **Nut splitting**

   Areca palms in the age group of 10-25 years are more prone to this disorder. This is prevalent in paddy field converted lands as well as gardens with high water table and seen during rainy season. Sudden flush of water after a long period of drought is the main cause. Proper drainage and application of potash fertilizers and spraying of Borax 0.2% in the early stage reduces the splitting.

3. **Band disease**

   Affected palms exhibit small crinkled dark green leaves, tapering stem and reduced internodal length. Generally good drainage and better soil management practices, balanced application of organic and in organic fertilizers, macro and micro nutrients reduce this problem.

CHAPTER

16

Palmyrah

PALMYRAH (*Borassus flabellifer* L.)

Family: **Palmae**

Palmyrah is believed to be a native of tropical Africa but grows extensively in the drier parts of India, Srilanka, Burma, Thailand, Vietnam, and Malaysia and in some parts of Indonesia. Many earlier historical writings did mention about the palm and its various uses. All the parts of the palm are useful and over 800 various uses are reported and hence this palm is known as 'Kalpagathara' i.e. Tree of life. Signifying its importance, Government of Tamil Nadu recognized it as the State tree of Tamil Nadu since 1978.

It has been estimated that there are about 8.6 crores of palmyrah palms existing in India, of which, 5.02 crores palms are alone spread over Tamil Nadu and the remaining 3.58 crores palms are in Andhra Pradesh, West and East coasts In Tamil Nadu. This palm is spread over an area of about 22,500 ha in all the districts except Nilgiris. The erstwhile Nellai district has the maximum area under palmyrah (12,055 ha).

Botany

This palm is a perennial plant, often reaching a height of 12-14 m and sometimes, even 90 m. Trunk is arboreous, cylindrical, black in colour,

218

bearing a terminal crown of 30-40 large fan like leaves. It is dioecious, bearing axillary spadices. Male flowers are embedded in fingers whereas female flowers are sessile. Fruit is a drupe, often seen in 40-50 per bunch, and is normally three seeded (Figure 19).

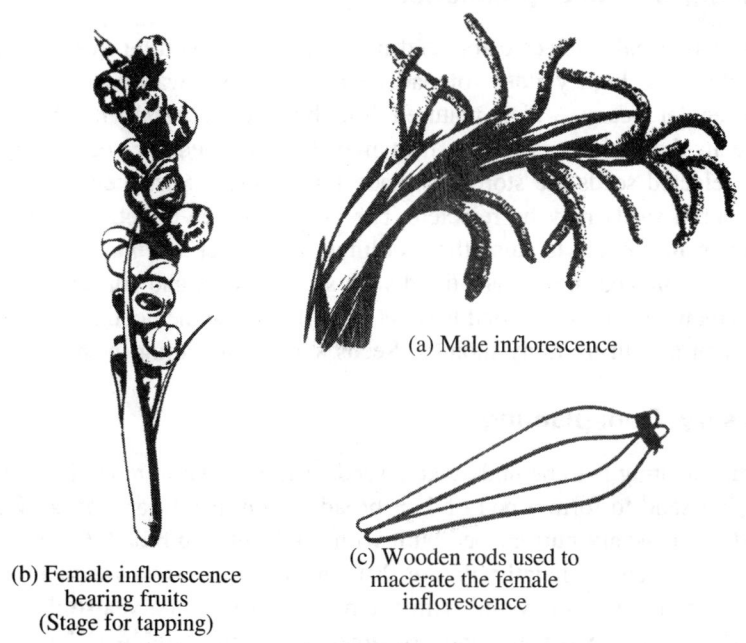

(a) Male inflorescence

(b) Female inflorescence
bearing fruits
(Stage for tapping)

(c) Wooden rods used to
macerate the female
inflorescence

Figure 19 Male and Female flowers in palmyrah.

Male and female palms can be distinguished only when they start flowering which occurs after 15-20 years of growth. Flowering season starts from January to August in Tamil Nadu but the peak season is May.

Climate and Soil

It can come up in dry areas having moderate to low rainfall. It grows from sea level up to an altitude of 750 m. Palmyrah can be grown in soils of all descriptions but found to grow well in deep, sandy and loamy soils.

Varieties

There is no recognized variety. They are generally classified as **Black skin** and **Red skin** fruited types. Palmyrah Research Station, Srivilliputhur (T.N.A.U) has released one improved variety viz SVPR-1 Palmyrah palm. It is a semi dwarf type with a high padaneer yield of 298 litres per palm in a tapping duration of 95 days. The padaneer of this variety has high jaggery content (144 g per litre of padaneer) and high brix content.

Raising of Planting Material

The mother palm selected for seed nut collection should be free from pests and diseases, high yielders of padaneer and fruits, dwarf in stature, early and regular bearers. Well matured fruit bunches indicated by the yellow tinge in the stylar end of the fruit may be harvested for seed collection. The selected seeds are stored in shade for 3 weeks. Shrunken, weight less and bored seeds may be rejected. The seeds may be directly sown *in-situ* or sown in nursery to raise the seedlings. For direct sowing, 3 to 4 whole fruits are planted in pits half filled with sand or sand and soil mixture. The pit is then covered with dried leaves (mulch) to the same depth. Sowing may be taken up during rainy months. Seeds start germinating in three weeks.

Nursery Transplanting

If transplanting is to be undertaken, seeds can be sown in mound formed by heaping sand to form a bed of 1 m broad, 30 cm height and of convenient length or masonry nursery bed built with bricks of 2 m broad, 60 cm height and of convenient length (Figure 20). In both the cases, seeds are sown at 10 cm space and covered with 5 cm sand. About one year old seedling is lifted from the nursery and containerized in the polythene bags. After roots are formed the seedlings can be successfully transplanted in the main field. Shading the seedlings should be provided using plaited coconut or palmyrah fronds.

Manuring

Farmers generally adopt sheep penning for their palmyrah plants since the soils are usually sub fertile. Application of 10 kg of FYM/pit before planting the nuts (as basal dressing) is advantageous. The dosage may be increased biannually till reaching 60 kg FYM/pit/year.

After Cultivation

After cultivation consists of gap filling, interploughing, basin rectification in the initial few years, defoliation and tending. Gap filling may be carried out using containerized seedlings. One ploughing during summer and another at the onset of North East monsoon may be given in the

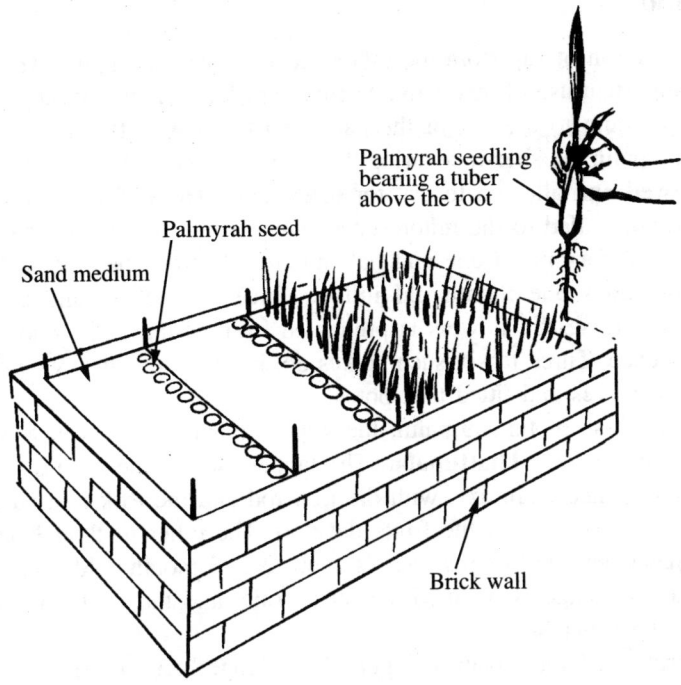

Figure 20   Masonry nursery in palmyrah.

interspaces. Basin rectification before rains helps for efficient collection and storage of rainwater. Tending consists of removing the persisting leaf bases periodically. Intercrops like cowpea, moringa, green gram, red gram, bengal gram and semiarid zone fruit crops like ber, amla, pomegranate, Carissa, West Indian cherry and guava appear to be good.

Growth and Yield

The first frond appears in about five months. Production of fronds is slow to begin with. However the pace picks up as the age advances. The first fan shaped tree leaves appears only in the second year. It will be 6-7 years before the top of the trunk appears. This palm is a slow grower and grows to a height of 12-18 m and comes to flowering in about 13-15 years (irrigated condition) when it can be tapped for padaneer. On an average, 100-200 litres of padaneer can be obtained for a period of 4 months from February to May. However the padaneer and fruit yield are highly variable in individual palms.

Tapping

The extraction of sap from the inflorescence is called tapping which is the most important use of this palm. There are different kinds of tapping which vary according to the sex of the palm and the age of the infloresence. In the male palm the sheath covering the young inflorescence (two weeks old) is removed and allowed to dry for three days. The end is cut every time and the pot is tied to the infloresence. This method is called **aripanai** and is practiced for one to one and half months. In the other method, known as **Vallupanai** one month old inflorescence is selected and each male spike bearing sessile flowers is pretreated by pressing and stroking of the inflorescence. Three to six such spikes are brought together, wrapped with palmyrah leaves and fitted to a pot.

In the case of female palm, the young female inflorescence is tapped by a method called **Thattupalai**. The tappers soften the tissues by hitting the inflorescence main axis with the iron rod and the fork is used to press the regions from which the fruits develop. The other method **Kaivetty** is employed when the inflorescence is about 2 to 3 months old. The inflorescence at this stage is matured and bear fruits and the fruits are sliced as the tapping progress.

Normally, female palm is tapped for a longer period (April to December) compared to the male palms (December-February). The sap is collected twice a day, in the early morning and in the evening. Each time, at the end of the collection of the sap, a new surface is made by cutting a thin slice of the inflorescence. During the collection of the sap, the climber takes a bigger pot and empties into it the sap collected in the small pots placed on each inflorescence. Hence, evening time when sap is collected, the bottom sediment which contains the yeast and bacteria are left behind. On every fifteenth day, the entire sediment is completely removed but not washed.

The sap so tapped, is called, **Neera** or **padaneer** which is transparent, pleasant smelling and sweet. It contains sugar (12-16 per cent), most of the essential amino acids and vitamins like ascorbic acid and B complex. Hence it is considered to possess medicinal properties too. This neera is drunk as such or used for preparation of secondary products through some processing.

1. **Toddy** is obtained as a result of natural uncontrolled fermentation of the sugary sap by the wild yeasts and bacteria which come in contact with the sap. The alcohol content in a fully fermented toddy is about 5 per cent.

2. **Jaggery** is a solid mixture of reducing and non-reducing sugars prepared by concentration of the strained palmyrah sap through gradual boiling.

3. **Palm sugar** is obtained by boiling the strained and clarified sap to 108-110°C to attain a saturation stage of 85° Brix and allowing it to crystallise.

4. **Palm cola** is the aerated soft drink containing 11 percent palm sugar having cola concentrate, citric acid and food colours as other ingredients.

Besides the above from neera, nungu and palmyrah tubers are also used. The tender palmyrah fruit is called nungu and is delicious and rich in corbohydrates, phosphorous, iron, vitamin C and riboflavins and niacin. Three months old roots (tubers) are edible on cooking by direct fire or boiled in water with salt.

Plant Protection

As in coconut-, rhinoceros beetle and black headed cater pillar often cause damage. Similarly, bud rot is often noticed in rainy months. Control measures are similar to those recommended for coconut.

CHAPTER

17

Oil Palm

OIL PALM (*Elaeis guineensis* JACQ.)

Family : **Palmae**

The oil palm is recognized as the cheapest edible oil yielding crop and has become one of the major oil crops in the world. The most productive oil bearing crop, oil palm yields about three times the oil yield of coconut, seven times that of rapeseed, and ten times that of soybean. Centre of origin of oil palm is tropical rain forest region of Guinea coast of West Africa. Total world production of palm oil is 45.85 million tonnes. Indonesia is the world's largest producer (22.22 million tones) followed by Malaysia (16.99 million tonnes). The crop is unique in producing two distinct types of oil –'palm oil' and 'plam kernel oil' used for culinary and industrial purposes. Being rich in β –carotene, palm oil contributes substantially to nutritional and energy requirements of masses.

Oil palm was introduced in India at National Botanic Garden, Kolkata in 1886; commercial cultivation of the crop was started in India in a big way during 1971-1984 in Kerala. In India, 7.96 million ha of land is identified as potential area for cultivation under assured irrigation, however, the present area is only about 1.64 lakhs ha, producing about 20 tones of fresh fruit bunches of oil palm per ha/annum. Andhra Pradesh, Karnataka, Tamil Nadu and Kerala are principal oil palm growing states in the Country. Gujarat, Goa, Maharashtra, Orissa, Assam, Tripura and West Bengal are also being identified to grow oil palm on a large scale as India is spending around Rs.3000 crores annually to import palm oil and vegetable oil.

Botany

Genus *Elaeis* belongs to family Arecaceae. It is an unbranched monoecious tree growing to a height of 20-30 m and which lives up to 200 years. Flowering begins early when its young palm is well established. An inflorescence primordium is produced in axil of each leaf at a time of leaf initiation. Inflorescence reaches central spear stage in 2 years and further 9-10 months are required for flowering and anthesis. Each flower primordium is a potential producer of male and female organs but one or the other usually remains rudimentary to produce either a male or female inflorescence. Hermaphrodite flowers and inflorescence are occasionally produced. The inflorescence, a compound spike or spadix has a stout peduncle, 30-45 cm long, and a central rachis with spirally arranged spikelets 100-200 in number in different palms.

Male inflorescence has finger- like spineless, cylindrical spikes, 10-20 cm long each with 700-1200 closely packed small male flowers sunk in tissues of rachis.Female inflorescence has thick and fleshy spikelets in axils of spiny bract and with terminal spine of varying length. Female flower has 2 bracteoles, 6 sepaloid tepals in 2 whorls, rudimentary androecium of 6-10 short projections, tricarpellary ovary, of which one carpel develops, and sessile trilobed stigma.

Climate and Soil

Oil palm requires evenly distributed rainfall of 150 mm per month or 2500-4000 mm per annum. If rain-fall distribution is not even and adequate in potential areas for oil palm cultivation in India, it can be grown in assured irrigated conditions by adopting recommended practices. It comes up well between 29-33°C maximum and 22-24°C minimum temperature. It enjoys a bright sunlight for atleast 5 hrs per day as well as high humidity of more than 80 %.

Oil palm comes up well in almost all types of soils but the best suited soils are well drained deep, loamy alluvial soils, rich in organic matter with adequate water holding capacity. Highly alkaline or saline, water logged and coastal sandy soils are to be avoided.

Variety

Cultivars in strict sense do not occur. The best classification based on fruit structures is as follows:

- **Dura:** Shell usually 2-8 mm thick, low to medium mesocarp content (35-55 %), kernels large, no fibre ring. In Deli dura palms, kernels tend to be larger, comprising 7-20 % of weight of fruits.

- **Pisifera:** Shell less with small pea- like kernels in fertile fruits. It is of little commercial value, but is important in breeding commercial palms

- **Tenera:** It is the ruling hybrid grown all over the world and is a cross between thick-shelled dura and shell less pisifera. Tenera has thin shell (0.5 to 4 mm thick), medium to high mesocarp content (60-90 %) and high oil content. It has higher sex ratio and larger number of bunches than dura.

Seed Collection

Tenera hybrids which are a cross between dura and pisifera are used for commercial planting. Elite dura and pisifera palms are selected based on progeny trials. Male and female inflorescences are bagged, 7 days prior to flower opening. A female flower is receptive when lobes of the stigma are well separated and pink. Receptive period of flowers in an inflorescence lasts for 36-48 hours. Therefore, if pollination is carried out on 3 successive days, maximum fertilization is obtained. Pollen collected from desired male parents, either fresh or stored, can be used for pollination by smearing on stigma or using a puffer. Pollination is normally carried out in morning. Bag is retained on bunch for 3-4 weeks after pollination. Bunches mature within 5-6 months and are harvested for seed purpose. Bunch is cut into spikelets and fruits are taken off by hand. Seeds are extracted by scrapping off the exocarp and mesocarp with a knife or retting in water.

Nursery

Polybag nursery is preferred to conventional field nursery. Polybags (preferably black) of 400-500 guage measuring 40 cm x 35 cm are used. Bags are filled with potting mixture and are arranged at a spacing of 45 cm × 45 cm. One healthy sprout is dibbled per bag. A good mulching during summer is desirable. Seedlings are watered thrice a week. A fertilizer mixture containing 15 g N, 15 g P and 6 g K @ 8 g/5 L of water for 100 seedlings may be applied from 2 to 8 months stage.

Planting

Planting can be done in any season. Best period is June to December coinciding with monsoon. Mulching and growing cover crops like sunhemp in the oil palm basin (1.5 meter radius) in two to three rows will help to overcome hot wind waves during summer. Twelve to fourteen months old healthy seedlings with 1-1.3 m height from base and 13 functional leaves with good girth at collar are used for planting. A distance of 9 m × 9 m × 9 m (triangular planting pattern) accommodating 143 plants per hectare can be followed. The pit size of 60 cm × 60 cm × 60 cm is normally dug one month prior to planting. Before planting, 400 g Single Super Phosphate or 250 g Rock phosphate and 50g Phorate are applied and mixed with the soil at the base of the pit. While planting, care should be taken to see that the seedling bowl will be 25 cm below the ground level. In case of low-lying wetland soils, planting should be done in raised mounds to avoid water logging and poor aeration.

Irrigation

Oil Palm requires sufficient irrigation, as it is a fast growing crop with high productivity and biomass production. Insufficient irrigation leads to reduction in the rate of leaf production, affects the sex ratio and results in inflorescence abortion and yield reduction. Each grown up yielding palms of 3 years age and above requires 200–250 liters of water per day. However, in older plantations during hot summer, 300–350 liters water may be required. If irrigation water is not a constraint, basin irrigation can be taken up. Irrigation channels are prepared in such a way that the individual palms are connected separately by sub-channels. Required quantity of water can be given at weekly intervals. For light soils frequent irrigation with less water has to be given. If more water is given at a time, leaching of nutrients will be more. In heavy soils, irrigation intervals can be longer. If irrigation water is limited and land is of undulated terrain, drip or micro sprinkler irrigation can be advantageous. In such case, 300-350 litres of water should be ensured during hot summer.

Manuring

Oil Palm is a gross feeder and demands a balanced and adequate supply of macro, secondary and micronutrients for growth and yield. The following fertilizer schedule is recommended, but efforts must be made to modify this dose based on soil and leaf nutrient analysis.

Age of the plant	Nutrient requirement/ palm/year (g)			
	N	P	K	MgSO$_4$
I year	400	200	400	125
II year	800	400	800	250
III year	1200	600	1200	500

For the newly planted crop the first dose of fertilizer application may be given three months after planting while from second year , the fertilizers are to be applied at every three months interval under irrigated conditions (in 4 splits). 50-100 kg FYM or 100 kg green manure per palm are applied along with the second dose of fertilizer application. Five kilograms of neem cake/ palm/year can also be applied. Whenever organic fertilizers are applied nitrogen application (through chemical fertilizers) can be reduced proportionately.

In gardens where 20 to 25 tonnes FFB/ha is obtained by 5th or 6th year, an additional 20 % of the recommended dose of fertilizers can be applied to maintain the productivity. If productivity levels are still high, an additional dose of one to two kilograms of muriate of potash per palm per year can be applied. Borax at the rate of 100 g/palm/year is also recommended when the deficiency symptoms are noticed. The fertilizers are broadcast around the clean-weeded basin, about 50 cm away from the palm base and incorporated into the soil by forking.

Basin Management

Basin area of oil palm represents active root zone, hence it must be kept clean and weed free to avoid competition for water and nutrients. During first year basins of 1 m radius is to be taken around the palm removing the soil from inside, subsequently, it can be widened to 2 m radius during 2nd year and to 3 m from 3rd year onwards. Mulching of oil palm base is essential for conserving moisture as well as to control weeds. Mulching can be done with dried leaves, male inflorescence, coconut husk; empty bunches brought from factory etc. In adult plantations all the cut leaves can be heaped in between two rows of oil palm, which can act as mulch. Regular manual weeding or chemical weeding is recommended to check the weed growth. Glyphosate @ 750 ml/ha/year or 17.5 ml/basin is recommended for effective weed control.

Intercrops

Oil palm is a wide spaced perennial crop with a long juvenile period of 3 years. The interspace during the first 3 years can be used to grow intercrops and generate income. The most suitable crops are vegetables, dwarf banana, maize, tobacco, chilli, turmeric, ginger, pineapple, flowers, etc. While raising intercrops avoid tying of palm fronds, which will reduce photosynthetic activity and ploughing close to the palm base, which will cut the absorbing roots and thereby reduce intake of water and nutrients. In mature oil palm gardens of 8-10 years age or palms attained a height of 15 feet; cocoa, pepper etc can be grown as intercrops. Cover cropping and green manure cropping in the alley spaces are encouraged for moisture conservation, improve soil condition, organic matter and also to inhibit weed growth.

Flowering

Oil Palm comes to flowering 14-18 months after planting. It produces both male and female flowers separately on the same palm. Male and female phases do occur naturally in a consequent cycles in a palm. Some individual trees may exhibit a phenomenon of producing more male inflorescence and less number of female inflorescence, however for normal yield annual production of 10-12 female flowers is satisfactory. Large number of male flowers occur due to:

(a) Insufficient irrigation and irrigation at longer intervals.

(b) Non-application of recommended doses of fertilizers in appropriate quantities at right time.

(c) Excessive pruning of fronds.

(d) Ploughing deeply and close to the palms damaging the active feeding roots.

The flowers are to be removed (soon after appearing) easily by hand pulling (i.e. ablation) up to 3 years. This enables the plant to gain adequate stem girth, vigour and develop adequate root system. Oil Palm is a highly cross pollinated crop. Pollination is assisted by wind and insects. Release of the weevil, *Elaeidobius kamerunicus* after 2 1/2 years helps in good pollination and fruit set.

Harvesting

Harvesting an important operation, which determines the quality and quantity of oil to a great extent. The maturity symptoms include (a) fruits in the bunch turn yellowish orange and (b) 5-10 fruits from each bunch drop on their own and (c) when pressed hard with the fingers orange yellow coloured oil exudes from the fruit. If unripe bunches are harvested oil content is less and recovery will be less. While harvesting, the stalk length would be less than 5 cm. In young plantations, more bunches with less bunches weight would be available while in adult plantations the bunch weight is more but the bunch number is less. Sharp chisel attached to aluminum or iron rod or harvesting knife fitted to aluminum pole extendable up to 45 feet are normally used to harvest the bunches.

Plant Protection

Important pests and diseases affecting the oil palm are alone indicated below:

Pest	Symptoms	Management
Rhinoceros beetle	V" shaped gaps in the frond. Hole on the petiole with chewed up fibre at that place.	1. Farmyard manure pits in the plantation should be treated with carbaryl 50 WP @ 0.01 % (200 g/1 ton of FYM) 2. Adult beetles can be trapped and killed using fermented castor seed/ cake bait or pheromone traps, 3. It can be controlled biologically with green muscardine fungus *Metarhizium anisopliae*. The fungal pathogen can be applied on breeding sites like FYM pits, dead logs of oil palm and coconut. 4. Keep 20 g of Phorate in perforated polythene sachet in spear leaf axis. Change the position of sachet every month by keeping it in a new spindle.
Red palm weevil	Palms show gradual wilting and drying, presence of few holes and oozing of brown viscous liquid from these holes at the base of the palm. Grubs feeding inside the trunk make characteristic sound.	Root feeding of Monocrotophos 10 ml in 10 ml of water.

Disease Management

Disease	Symptoms	Management
Bud rot	(a) Yellowing of the spear leaves which subsequently turn to brown. (b) Affected spear bends at the base and seen hanging down in the crown. (c) The basal tissues of the spear completely gets rotten as a result it collapses and can easily pulled out.	The affected spear should be pulled out along with the decayed tissues. The affected tissues of the meristem should be removed layer by layer till fresh tissues are seen and carbendazim paste (mixing 20 g in 100 ml of water) applied.
Basal stem rot (*Gonoderma spp.*)	(a) Withering, yellowing and orange discoloration of the leaves followed by necrosis on one side of older fronds. (b) Desiccated fronds drop or break at some point along the rachis. (c) Appearance of light brown lesions or rotting of the bole at the stem base is characteristic symptom at the advanced stage of disease. The infected palms appear suffering with malnutrition. (d) The disease produces dry rot of internal tissues at the base of trunk.	(a) Removal and destruction of the dead and diseased palms. The palms in the early or middle stages of the disease should be isolated from the neighbouring palms by taking circular trenches of 1 meter depth and 30 cm width at 3meters diameter from the palm base. (b) Affected palms should be applied with 5 kg of Neem cake per year. (c) The disease affected and surrounding palms should be given with Calixin 10 ml in 100 ml of water as root feeding stem injection.
Stem wet rot	(a) Sudden death of spear leaves including the young expanded fronds surrounding the spear. (b) Remaining fronds show yellowish discoloration and then rapidly wither and die. (c) Sometimes, the older fronds die first and the symptoms progress to the young ones. (d) For confirmation of disease a sharp iron rod/ auger may be pierced into the	(a) Burn the dead and diseased palm after removing from the plantation. Improvements in agronomic practices, providing drainage, avoid flooding of the field etc., (b) Early detection of the disease and trunk surgery can save the palm. **Trunk surgery** Remove the rotten tissues completely. When the surgery

	stem base on four sides, which gives out liquid with putrefied smell.	is completed, a protective covering with carbendazim 1% mixed with monocrotophos paste followed by hot coal tar layer is given. Repeat the surgery after 3 months if symptoms reoccur.
Spear rot (Associated with Mycoplasma)	Chlorosis of younger leaves followed by rotting of spear, shortening of the leaves and leaflets and root dacay are noticed. Bunch production is reduced and infected palms die in 3-5 years.	Roughing of diseased palms.

Pollinating Weevil

Oil Palm is an entomophilous crop. The major cause of bunch failure in oil palm is the paucity of pollen. The introduction of *Elaeidobius kamerunicus* is expected to raise levels of pollination and fruit set in oil palm. The weevil is much better adapted to oil palm and are equally effective pollinators on young and old palms. Weevils are highly specialized and are also less affected by adverse weather and soil conditions. They breed on spent male inflorescences and visit female inflorescences only during anthesis. They do not injure the female flowers or the fruit. *Elaeidobius* species visit both male and female inflorescences of oil palm. It is observed that the limit of searching ability is about 1000 m on a dry comparatively breezeless day. The adults of the beetle are attracted to female flowers. The adults visit female flowers but cause no injury. It therefore appears to be the most promising candidate for introduction into Oil Palm plantations having pollination problem.

Postharvest Handling

Rapid extraction of fruit from bunch and sterilization of bunch are important, since the complete natural loosening of fruits takes a week or more, resulting in increase in free fatty acids (FFA) lowering the oil quality. Steam sterilization of fruit bunches at 3 kg/cm^2 pressure for 30-60 minutes is done to arrest development of FFA and also to soften and loosen fruits for easy stripping and pounding. Sterilizers with 1-10 tonnes capacity are utilized for this purpose. Sterilized fruits are separated from bunch by passing through a rotary drum stripper. Stripped fruits are converted to a homogenous oily mass by pounding and digested by heating during which cell walls are ruptured releasing entrapped oil. Pounded mass is squeezed under hydraulic press which separates palm oil from fibre and seed. Oil is boiled in a clarification drum and sludge containing water and fruit solids settle at the bottom and oil floating on the surface is drained out. Oil is passed through centrifugal purifier to remove remaining solids and vacuum dried to remove moisture. It is then pumped to storage tanks and stored as crude palm oil. Crude palm oil from clarifier is passed through a high speed centrifuge at 80^0C to remove traces of solid impurities and water. Final traces of moisture are removed by vacuum drying and pure raw palm oil is stored in tanks.

Processing of palm kernel oil involves the deoiled fibre/ nut mixture after pressing passes a pneumatic separation system. The fibre after separation is used as a fuel. The nuts are cracked using centrifugal crackers and separated into kernel and shell using air and water separator system. The shell is used as a fuel and kernel is dried to a moisture content of 6-8 %. The kernel is powdered and steam conditioned followed by expression of oil in expeller. The palm kernel oil recovery is about 2 % of FFB.

CHAPTER

18

Cinchona

CINCHONA (*Cinchona* sp.)

Family: **Rubiaceae**

Cinchona is native of high lands of South America and was introduced in India (Nilgiris) in 1859. Now, it is grown in few ha in Nilgiris and Anamalai hills of Tamil Nadu. It is also grown in Darjeeling (West Bengal).

It is an evergreen tree, growing to, a height of 10-12 m with a sparse branching habit. The important species which are under commercial cultivation are *C.ledgeriana, C.officinalis, C.robusta* and *C.succiruba.* As a result of natural cross pollination, the present day plantations are admixtures of the various species.

'Quinidine', an alkaloid of cinchona bark is used for its anti-malarial, anti-pyretic and oxytonic actions. But no clinical use is currently made of these properties. In addition to their use in pharmacy, quinine and quinidine and their derivatives are utilized in insecticide compositions for the preservation of fur, feathers, wool and textiles.

Climate and Soil

Cinchona requires an average temperature of 20° with a relative humidity of 85 per cent. Annual rainfall should be not less than 200 cm but distributed

234

atleast 8 months in a year. The best elevation is 1000 to 2000 m above M.S.L. without any frost occurrence. Cinchona prefers porous, well drained, fertile soils with a thick cover of organic matter and high moisture holding capacity. The optimum pH range is 4.5 to 6.5.

Propagation

Cinchona is propagated by seeds and vegetatively by cutting, stooling, layering and patch budding. Seeds are sown in raised beds during April and they take about 20-30 days for germination. The healthy seedlings are transplanted in baskets or polythene bags when they are about four months old. Clonal propagation is sometimes done through top working or patch, budding.

Planting

The area selected for planting should be cleared a year in advance and planted with shade trees like Silver oak and or dadabs. Pits of 30 × 30 × 45 cm are dug and filled up with top soil and other well decomposed organic matter. Seedlings are transplanted in the main field at a spacing of 1.25 × 1.25 m when they are about one year old. Transplanting is done any time when there is sufficient moisture in the field. The other method is to go for high density planting ie. trees are set out at a spacing of 1.0 × 1.25 m or 8000 plants per hectare and gradually harvested until 800 plants/ ha remain after 25 years.

Manuring: Annually, cinchona plants are manured with 115 kg N, 15 kg P_2O_5 and 115 g K_2O ha. Once in 3 to 4 years when the soil pH goes below 4.5, liming @ 1.0 to 1.5 t/ha is recommended.

After culture: The main cultural operations are staking the plants in the first three years to prevent wind damage. In young plantations, wind growth may be checked by slash weeding, followed by chemical weeding with paraquot @ 30 ml and sodium salts of 2, 4-D @ 25 g in 10 litres of water.

Harvesting

The trees are coppiced when they are 8 to 10 years old depending on the vegetative growth. Coppicing consists of pruning the tree at a height of 5 cm from the ground level. The stump left out regenerates to produce a large number of shoots but only two or three of these are retained and allowed to grow further while the remaining ones are removed. The trees become ready for the second coppicing within 8 to 10 years from the time of first coppicing. After the second coppicing also, two to three shoots are left to grow further. The trees are finally uprooted in the 30th year when they start declining in vigour. Even though the major harvests are obtained at the time of first two coppicing, some yields of bark are also available from the dead and dying trees and prunings. During the first two coppicing, a yield of 4000 kg of dry stem bark per hectare may be obtained and at the final stage of uprooting the tree, the yield of bark may be 6000 kg per hectare. The most important alkaloid principle is Quinine which, occurs in the stem, twig and root bark of the tree. Normally its content range from 3 to 4 per cent as Hydroxy Quinine sulphate.

Processing

The extraction of quinine involves beating of the bark with a mallet to loosen it for peeling by hand or knife. The peeled bark is quickly dried to prevent the loss of alkaloids. The fully dried bark is sent to the factories for solvent extraction of powdered bark with slaked lime containing more than 60 % of Calcium hydroxide and the alkaloids removed with amyl alcohol or ether. These alkaloids are in turn extracted from the solvents in acidified water; they precipitate out when the water is made alkaline. It is then dried and powdered, and is the starting material for the manufacture of quinine base and other quinine salts. Medicinally, cinchona alkaloids form one of the most important groups of compounds. They have been administered in the form of extract, tincture, alkaloid mixture or as isolated alkaloids (purified) and their salts conforming to the latest pharmaceutical standards in various countries.

Plant Protection

In the nursery, damping off caused by *Pythium* is often noticed. Drenching with 0.5% copper oxy chloride is recommended at 10 days interval.

Tea mosquito bugs (*Helopeltis antonii*) often infest the leaves in the nurseries and also in the main field. Spraying with any systemic insecticide will check the incidence.

PART-III
MEDICINAL PLANTS

CHAPTER

19

Introduction

Medicinal plants are those plants rich in secondary metabolites and, are potential sources of drugs. These secondary metabolites include alkaloids, glycosides, coumarins, flavonoids, steroids etc.

These plants form the main base for the manufacture of drugs of Indian Systems of Medicine (ISM) (Ayurveda, Unani, and Siddha) and Homeopathy. These plants are found in various parts of the country in different environmental and climatic conditions. Plants which grow wild in forest regions, classified as **minor forest produce**, supply a substantial amount of raw material required for the indigenous drug industry.

Importance and scope for cultivation of medicinal plants in India

1. The ancient Indian System of Medicine (ISM) is predominantly a plant-based materia medica making use of most of our native plants. It caters to almost the entire rural population of our country mainly because of the scarcity of modern allopathic health care in our villages.

2. World Health Organisation (WHO) has listed over 21,000 plant species used around the world for medicinal purpose and in India 8000 species listed is available. ISM uses nearly 2500 plant species belonging to more than 1000 genera. Of the nearly 800 species used in the industry, 25 % are cultivated.

3. India is one of the few countries where almost all the known medicinal plants can be cultivated in some part of the country or the other. Among the various plants in great demand in the country and abroad are opium poppy, tropane alkaloid bearing plants, sapogenin bearing yams, senna, psyllium husk and seeds, cinchona and ipecac.

4. ISM offers most appropriate or first line therapy against many diseases like jaundice, bronchial asthma, rheumatoid arthritis, diabetes etc. for which allopathic medicines have as yet no cure. It is well known that most allopathic medicines produce many morbid side-effects. It is for this reason that more and more people in the western societies are showing increasing interest and preference for organic drugs and their preparations.

5. India has a vast geographical area with high production potential and varied agro- climatic conditions that it is possible to grow any medicinal plant. Most of these plants can subsist under stress conditions and are thus suited even for rainfed agriculture. Cultivation of medicinal plants offers considerable scope for rural employment and export for foreign-exchange earnings.

6. India has nearly 9500 registered herbal industries and a multitude of unregistered cottage level herbal units which depend on the continuous supply of medicinal plants for manufacture of herbal medical formulations based on ISM.

7. India is already a major exporter of medicinal plants. It is estimated that nearly US $ 53 million worth of raw materials and drugs from medicinal plants are exported from India. It holds monopoly in the production and export of psyllium and senna and is the second largest exporter of opium latex.

8. Though India ranks in number two in terms of export of medicinal plants, it occupies only sixth position in terms of value as we are exporting only raw material and not the finished drugs.

9. Many of the medicinal plants required by the trade are gathered mainly from the wild growth (nearly 80 %) thus depleting the vegetation of its valuable medicinal plant wealth (eg. Rauvolvia, Dioscorea). On account of this practice, many species of medicinal plants in our country have become extinct or endangered. This should be prevented and herbal gardens and gene-banks covering important medicinal plants should be established to conserve them. Deforestation, shifting cultivation, over grazing and over exploitation are the other main human activities posing a threat to biodiversity and medicinal plant wealth in this country. Ministry of Commerce,

GOI has banned export of 52 medicinal plants in order to conserve them in nature and semi-wild conditions.

10. World Conservation Union (IUCN) has identified following categories:

Category	Brief Description	Plants to this category
Extinct	The last remaining member of the species has died or is presumed beyond reasonable doubt to have been dead.	*Myristica malabarica, Rauvolfia serpentina*
Extinct in the wild	Captive individuals survive, but there is no natural population	*Plectranthus vetiveroides*
Critically endangered	Faces an extremely high risk of extinction in the immediate future	*Adhatoda beddomei, Janakia arayalpathra*
Endangered	Faces high risk of extinction in the immediate future	*Nilgirianthus ciliatus, Nervilia aragoana*
Vulnerable	Faces high risk of extinction in the medium term	*Garcinia indica, Adenia hondala*
Rare	Not available plenty in nature	*Vernonia anthelmintica, Commiphora mukul*

11. This warrants conservation and sustainability of the plant species for present and future generations. One approach is resorting to *in-situ* conservation by establishing bioreserves, national parks, wild life sanctuaries, sacred grooves and other protected areas (already India has 4.5 % of its total geographical area under this category). The second approach is to establish *ex-situ* conservation i.e. cultivating and maintaining these plants outside their natural habitat in botanical gardens, institute gardens, gardens maintained by NGOs etc.

As the medicinal plants are very many in number, it is difficult to cover all the plants in this part of the book. The cultivation details of selected commercially important medicinal plants are dealt in chapter 20 while a brief description of about 40 important medicinal plants are mentioned in chapter 21. A glossary of some medical terms used in the text are given in Annexure-I.

CHAPTER

20

Commercial Medicinal Plants

20.1. MEDICINAL YAM
(*Dioscorea floribunda* Mart and Gal)

Family: **Dioscoreaceae**

Diosgenin, obtained from Dioscorea tubers, is the major base chemical for several steroid hormones including sex hormones, cortisone, and other corticosteroids and is the active ingredient in the oral contraceptive pill. The growing need for steroidal drugs and the high cost of obtaining them from animal sources led to a widespread search for plant sources of steroidal sapogenins, which ultimately led to the identification of the genus *Dioscorea* as the most promising one.

This genus *Dioscorea* belonging to family Dioscoreaceae with over 600 species is widely distributed in tropical world. Some of the species like *D. alata* and *D.esculenta* are under cultivation for long time for their edible tubers. There are about 15 species of this genus containing diosgenin. Among this, *D.floribunda* and *D.composita* are widely grown for diosgenin production.The major Dioscorea producing countries are Mexico, Guatamala, Costa Rica, India and China. The estimated world consumption of diosgenin is somewhere between 800 to 1000 tonnes per year. India produces only 25 tonnes of diosgenin annually mainly from the natural source and to be self sufficient, India has to step up its production to the tune of about 200 tonnes of Diosgenin per year.

The cultivation hints for *D.floribunda* and *D.composita* are discussed here:

D. floribunda

The best adapted species in Karnataka, Assam, Meghalaya, Tamil Nadu, Goa, & Andaman is *D.floribunda,* an introduction from Mexico. It produces compact tubers at a shallow depth. The diosgenin content varies from 2 to 7 per cent depending on the age of the tubers.

Climate and Soil

It is a tropical species preferring a tropical climate without extremity in temperature. It is adapted to moderate to heavy rainfall areas. Dioscorea plants can be grown in a wide variety of soils, but light soil is good as harvesting of tubers is easier in such soils. Extremely heavy clay soils are, in general not recommended, as they restrict tuber growth and make harvesting difficult. Dioscorea tolerates fairly wide variation in soil pH, though very acid soils should be avoided, the ideal soil pH being 5.5 to 6.5.

Varieties

The Indian Institute of Horticultural Research, Bangalore has released so far two improved varieties.

1. **FB (c)-l** - a composite strain of **D. floribunda,** which has been released for commercial cultivation. This is a vigorous growing strain relatively free from diseases. This has a diosgenin content of 2.5 to 3 per cent.

2. **Arka Upkar** - a high yielding clone, selection from FB(C)-1. The plants are very vigorous with a stout, robust vine, bearing broad dark green leaves. The tuber branches are thick, broad and deep. It has a higher diosgenin content of 3.5 to 4.0 per cent.

3. **Pusa–1:** Released from IARI, New Delhi, a selection from germplasm with higher tuber yield of 1.5 kg per plant after 18 months

4. **S-3** - It is released by CIMAP, Lucknow. High sapogenin content, diosgenin (17 %)

Propagation

Dioscorea floribunda can be propagated by tuber pieces, single node stem cuttings or seed. Commercial planting is normally established by tuber pieces only. Propagation through seed progeny is variable and it may take longer time to obtian tuber yields.

Propagation from tuber pieces is accomplished by cutting the tubers into pieces weighing about 50–70 g each. Three types of tuber pieces can be distinguished for propagation purpose, viz (1) Crown, (2) Median and (3) Tip. Crowns produce new shoots within 30 days after planting, since they have preformed buds. Medians and tips may take up to 100 days to sprout. Crowns are therefore preferred for commercial planting. However, if there is a shortage of material, median and tip portions can be used for planting. Dipping of tuber pieces for 5 minutes in 0.3 % solution of Benlate followed by dusting the cut ends with 0.3 % Benlate in talcum powder before planting or storage in moist sand beds effectively checks the tuber rot. This treatment is very essential for obtaining uniform stand of the crop. The best time for planting is by the end of April so that the new sprouts will grow vigorously during the rainy season commencing in June in India.

For rapid multiplication of the elite materials in the initial stages, single leaf node cuttings can be adopted. The cutting consists of a single leaf with petiole and about 0.8 cm of the stem. Such cuttings are prepared from non-flowering plants. They are pre treated with 500 ppm of IBA solution by quick dipping and are planted in the mist chamber in sand beds. Within 8–10 weeks, these cuttings are transferred to plastic bags containing equal mixtures of sand, soil and farm yard manure. They will be ready for transplanting in the main field in six months time. This method is not recommended for commercial planting as the growth is very slow but useful to initially multiply the elite materials in larger number.

For raising a crop from seed, fresh seed should be sown in 8 cm x 12 cm plastic bags in the month of February. Polythene bags may be filled up with a pot mixture containing equal parts of sand, soil and farm yard manure. Vermiculite should be used on the top. Atleast two well filled seeds may be sown in each plastic bag at a depth of 0.5 to 1.0 cm. The nursery may be protected from drying by light shade. The bags may be watered with care, lightly and frequently. The seeds complete germination in four weeks. The seedlings being vine should be supported promptly with thin twigs.

The best season for transplanting the seedlings in the fields is June-July. Vigorous seedlings may be alone transplanted and others may be discarded. The botom and sides of bag may be cut before transplanting so as to transplant the seedlings without disturbing the root system. As the progenies raised from seeds are highly variable and their growth is slow, this method is not recommended for commercial plantations.

Planting

Land should be prepared thoroughly till a fine tilth is obtained. Deep furrows should be made at 60 cm distance with the help of a plough. The stored tuber pieces which are ready for planting, seedlings or single node stem cuttings should be planted in furrows with 30 cm between the plants for one year crop and 45 cm between the plants for two years crop. The tuber pieces are planted at about 0.5 cm below the soil level. The new sprouts should be staked immediately. After sprouting is complete, the plants could be earthed up. Soil from the ridges may be used for earthing up so that the original furrows will become ridges and *vice versa*.

After Care

Dioscorea vines need support for their optimum growth and hence the vines are trained over pandal system or trellis.

Weeding: Initially, the vines are weak and tender and can not compete efficiently with the surrounding weeds. Periodic hand weeding, as and when necessary, is essential for the first few months. Experience has shown that once the plants have climbed up on the pandal, the weed population is considerably reduced due to shading. The plants by this stage can also compete more successfully with weeds.

Manures and Fertilizers: *D.floribunda* requires high organic matter for good tuber formation. Besides a basal dose of 18 to 20 tonnes of FYM per ha, a complete fertilizer dose of 300 kg nitrogen, 150 kg phosphorous and 150 kg of potassium should be applied per hectare. Phosphorous and potassium should be applied in two equal doses one after the establishment of the crop during May–June and the other during vigorous growth period of the crop (August–September).

Irrigation: Irrigation may be given at weekly intervals in the initial stage and afterwards at about 10 days interval. However during rainy season no irrigation is needed. In Anamalais, it is raised under pure rainfed conditions.

Interopping: Intercropping with legumes like cowpea, horse gram, cluster bean and French bean has been found to smother weeds and also provide an extra income without adversely affecting the tuber yield and diosgenin content.

Duration: The diosgenin content tends to increase with age; (2.5 to 3.0 per cent during the first year and 3.0 to 3.5 percent in the second year) and also the tuber yield and hence a two year crop is found to be more economical.

Harvesting

The tubers grow to about 25 to 30 cm depth and hence harvesting is done by manual labour. The best season for harvesting is Feb- March, coinciding with the dry period.

On an average 50 to 60 tonnes of fresh tubers can be obtained from one hectare in two years duration.

Plant Protection

The major pests of Dioscorea are the aphids and Red spider mites. Aphids occur more commonly on young seedlings and vines. They feed on the young leaves and stem. Young leaves and vine tips eventually die if aphids are not controlled. Older growth is seldom affected, Red spider mites attack the underside of the leaves at the base near the petiole. Severe infestations result in necrotic areas, which are often attacked by fungi. Both aphids and spider mites can be very easily controlled by spraying Abamectin @ 0.3 ml/L. No serious disease is reported to infect this crop.

D. composita

Another potential species of Dioscorea which offers a promising source of diosgenin is the **D. composita.** Due to the difficulties in its propagation, the area under this crop has not expanded and is now confined to few Government Farms in Nilgiris, Anamalais of Tamil Nadu, West Bengal and in Jammu.

This can be propagated by seeds, tubers or leaf node cuttings. Seed propagation is however commonly practised. March-April is the best season for raising seeds. Seeds are pretreated in cold water for 3–4 hours and then sown in pans containing equal parts of soil, sand and farm yard manure. Seeds start germinating within 2–3 weeks and complete germination by 4 weeks. When the seedlings are about 2–3 leaved stage, they are transplanted to polythene bags of 15 cm high, having shola soil, sand and farm yard manure. 3 to 4 months old seedlings can be transplanted to the main field at 90 × 60 cm distance.

Propagation through tuber is recommended for clonal multiplication of selected, high yielding types. But the tubers exhibt strong dormancy for a period of 2–3 months and thereafter erratic sprouting occurs. Maximum sprouting occurs if small tubers of one kg size are used for planting. Such tubers sprout within 90 days.

Supporting system for, *D. composita* is similar to that of *D. floribunda*. Vines may be allowed to grow for a minimum period of 3 years. Tubers will have a diosgenin content of about 3 %. Well maintained plant can yield 3–4 kg of tubers per plant.

20.2. FOX-GLOVE (*Digitalis lanata* Ehrh.)

Family: **Scrophularaceae.**

The genus *Digitalis* includes several species which are either medicinally or ornamentally important. *D. purpurea* and *D. lanata* are the two species recognised medicinally important as they contain three lanatoside A, B, C, which on hydrolysis yield digitoxin, gitoxin and digoxin, of which *digoxin* is the most important. The main use of these is in heart diseases and these glycosides have so far not been synthesised and hence cultivation of this crop is the only source of production of these glycosides. India requires about 3000 kg of this drug annually. The hill stations of India offer excellent scope for the cultivation of this crop.

Digitalis is native of Europe and is a tall herbaceous biennial or perennial plant. The leaves of *D. purpurea* is broadly lanceolate than that of *D. lanata* while the glycoside content is 2 to 3 times more in *D.* **lanata** (1 - 1.4 %). *D. purpurea* has 0.2 to 0.4 % glycoside content. *D. purpurea* produces white or purple coloured much elongated raceme while *D. lanata* produces relatively shorter inflorescence with smaller flowers having cream or yellow colours.

Climate and Soil

Digitalis does not perform well in places having extreme temperatures. It prefers a temperature range of 20° to 30°C for better growth and development. At lower temperature, vegetative growth of the plants as well as the glycoside synthesis is retarded. Altitude plays a critical role in growth and development and the best elevation is 1200 to 1800 m above MSL.

Digitalis requires a well drained sandy soil rich in organic matter. Soils with poor drainage are detrimental. It loves acidic soils, with a pH range of 5.5 to 6.5.

Cultivation

There is no improved variety available in India. One exotic introduction (D 76) available at Y. S. Parmar University of Horticulture and Forestry, Nauni Solan which can yield 39.12 q/ha of dry yield with a glycoside content of 0.87 % can be recommended. Digitalis can be raised from seeds either by direct seeding or by first raising in a nursery and then transplanting them to the main field.

The land should be thoroughly ploughed, leveled and weeded for direct seeding. The seeds may be broadcast or sown in lines. The second method is preferred as it facilitates subsequent inter cultural operations. About 8kg of seeds are required for sowing one hectare. Deep sowing below 2 cm should be avoided as it affects germination.

Another method is through raising nursery. The nursery beds are raised in well prepared places and the beds are sown in lines at 10 cm intervals and lightly covered with about 1 cm of fine sand. The seedlings will be ready for transplanting 40–45 days after sowing. Generally transplanting results in very high mortality of plants, therefore, direct seeding seems to be advantageous over nursery raising.

Spacing recommended is 45 cm between rows and 30 cm between plants in a row.

Digitalis may be applied with 25 to 30 tonnes of farmyard manure per hectare. Besides, 30 kg N and 8 kg P_2O_5 may be applied every year. In the first year the first dose may be given 2 months after germination or during February-March in the second year after the receipt of the summer showers. It also responds to foliar nutrition in the form of urea (1.0 %).

Harvesting and Processing

Leaves are the economic parts. Their age and time of picking affect the glycoside content. It is also observed that during noon the glycoside content is more and it is very poor during night hours. Under Indian conditions, 2 to 3 harvests of leaves can be made in the first year while in the second year 2 harvests besides a seed crop can be had. Therefore, it is recommended to keep the plantations as annuals with 2 to 3 harvests in the first year and in the second year, some portion of the first year crop should be earmarked for seed production and from this crop, leaves have to be harvested just before flowering.

Harvesting consists of plucking 8–10 cm long leaves without petioles. The interval between any two harvests can be 45 to 60 days Under Solan and Kodaikanal condition, the glycoside content ranges from 0.44 to 0.710 per cent.

Drying the leaves before storage at proper temperature is another important factor to retain the glycoside content. It is generally recommended to dry the harvested leaves by passing hot winds at 60°C with occasional stirring. The dried leaves should be stored in airtight containers with dehydrating substance like silica gel.

On an average 2 to 5.5 tonnes of dried leaves can be obtained from one hectare.

20.3. OPIUM (*Papaver somniferum* L.)

Family: **Papaveraceae**

The opium poppy *Papaver somniferum* is an outstanding medicinal plant, the products of which viz opium and codeine are important medicines used for their analgesic and hypnotic effects. A semi-synthetic derivative of this drug from morphine known as heroin has led to world-wide social problems. But attempts to find a synthetic drug which would replace morphine and codeine have not been fruitful so far. Its cultivation in India is confined to states of Madhya Pradesh, Rajasthan and Uttar Pradesh.

Papaver somniferum is an erect, rarely branched, glaucous annual, growing to a height of 60 to 120 cm. The leaves are ovate, oblong or linear oblong, flowers are large usually bluish with a purplish base or white, purple or variegated. It produces capsular type of fruits from which the latex known as opium is obtained on lancing. The fruits are about 2.5 cm in diameter, globose in shape. Seeds are reniform with white or black in colour. Though nearly all parts of the poppy plant contain white milky latex, the unripe capsules contain large amount.

Climate and Soil

It is a crop of temperate climate but can be grown successfully during winter in sub-tropical regions. Cool climate favours higher yield, while higher day/night temperature generally affects the yield. Frosty or desiccating temperature, cloudy or rainy weather tends to reduce not only the quantity but also the quality of opium.

Opium poppy prefers a well drained, highly fertile, light black or loam soil with an optimum pH around 7.0.

Varieties

A large number of races of opium known by their local names are reported to grow in India. They usually vary in leaf characters, floral characters or capsular characters. Telia, Dholia are some of the local races recommended for commercial cultivation. However, the following are the improved varieties available in Opium Poppy.

Variety	Breeding Method	Parents	Important traits	Institute/ University
Jawahar Aphim 16	Pure line selection from MOP 16	Selection from local bulk	Flowers: White serrated petal, Peduncle: Non hairy, Leaf: Long, slightly serrated and not waxy, Capsule shape: Spherical, flat at the top and surface groved	AICRP on M&AP, College of Agriculture, Mandsaur
Kirtiman	Selection from NOP 4	Local collections	Medium structure and high latex yield.	AICRP on M&AP, ND University of Agriculture and Technology, Faizabad
Jawahar Opium 539	Selection from MOP 539	Local collations	Plant: Medium height, Downy mildew resistant Flowers: White non serrated petal Peduncle: Non hairy Leaf: small, deeply serrated and non waxy and leaf lamina is narrow Capsule shape: Spherical, 2–3 capsule/plant, uneven surface	AICRP on M&AP, College of Agriculture, Mandsaur
Jawahar Opium 540	Selection from MOP 540	Local collations	Flowers: White non serrated petal Peduncle: Non hairy Leaf: Broad, slightly serrated and waxy Capsule shape: Oval, surface smooth, waxy	AICRP on M & AP, College of Agriculture, Mandsaur

| Chetak Aphim | Selection from EO 285 | Local collations | About 90 cm height, leaf is deeply serrated, flower initiation is bettern 31 and 37 days, have only one capsule per plant and white flowers; latex yield 52.7 and seed yield 1000 kg ha^{-1}. | AICRP on M&AP, Rajasthan Agricultural University, Udaipur |
| Trishna | Selection from IC –42 | Local collations | Flower colour pink; 5–7 capsules; morphin 14.78; latex yield 49–53 kg/ha | NBPGR, New Delhi |

Other Varieties released are:

Talia: Sown early, remains in field for 140 days. Flowers are pink in colour with large petals. The capsule is oblong, ovate, light green and shiny (waxy).

Ranghatak: It is a medium tall variety, mature for lancing in 125–140 days after sowing, white and light pink flowers, produces medium sized capsules which are slightly flattened on the top. Yields opium of comparatively thin consistency that changes to a dark brown colour on exposure.

Dhola Chota Gotia: It is a dwarf cultivar(85–90 cm), mature for lancing in 105–115 days and for seed in 140 days, purple white flowers and light green capsules which are oblong- ovate in shape.

MOP-3:
This variety developed at the Jawaharlal Neru Krishi Vishwa Vidyalaya, Mandsaur. It bears pinkish- white flowers with non –serrated petals, capsules are ready for lancing in 120 days after sowing.

MOP-16:
This variety developed at JNKVV, Mandsaur. It bears white flowers, serrated petals and round flat topped capsules. Drought tolerant, ready for lancing in 105-110 days after sowing. Recommended where early maturing crop is preferred.

Shama:
It is released by CIMAP, Lucknow , main alkaloids like Morphine (14.51 – 16.75 %), Codeine (2.05- 3.24 %), Thebaine (1.84–2.16 %), Papaverine (0.82 %) and Narcotine (5.94–6.5 %) are on higher side compared to other commercial varieties.

Shweta:
It is released by CIMAP, Lucknow, rich in main alkaloids like Morphine (15.75 – 22.38%), Codeine (2.15– 2.76%), Thebaine (2.04– 2.5%), Papaverine (0.82– 1.1%) and Narcotine (5.89–6.32%) .

Shubhra:
It is released from CIMAP, Lucknow for high morphine and seed yield.

Sampada:
Released by CIMAP, Lucknow, latex yield 80-90 kg/ ha, seed yield 0.7 to 0.8 tons/ha, resistant to downy mildew.

Sanchita:
It is released by CIMAP, Lucknow. It is concentrated poppy straw variety, morphine present in straw. Straw yield 0.7 to 0.8 t/ha, seed yield 0.8 to 0.9 tons/ha.

Vivek:
Released by CIMAP, Lucknow. It is a concentrated poppy straw variety, morphine present in straw. Straw yield 0.8 t/ha, seed yield 0.9 tons/ha.

Rakshith:
Released by CIMAP, Lucknow. It is a concentrated poppy straw variety, morphine present in straw. Straw yield 0.8 t/ha, seed yield 0.9 tons/ha. Resistant to downy mildew.

BROP-1 (Botanical Research Opium Poppy 1)
A synthetic variety developed at NBRI, Lucknow by crossing selections from Kali Dandi, Suryapankhi and Safed Dandi. Highly adaptable to varied agro- climatic conditions, higher yield, moderately resistant to diseases. Morphine content is 13 % and above.

Sujatha:
It is opium free poppy for production of oil and seed released from IIIM, Jammu.

Preparation of Land

The field should be ploughed 3 or 4 times to produce well pulverised soil. The field is then prepared into beds of convenient size.

Sowing: The seed is either sown broadcast or in lines. Before sowing, the seeds may be treated with fungicides like Dithane M.45 @ 4 g per kg of seeds. Seed is usually mixed with fine sand before broadcasting to ensure uniform spread in the bed. Line sowing is preferred to broadcasting as the latter method has many drawbacks like higher seed rate, poor crop stand and difficulty in carrying out inter cultural operations. The best time for sowing is late October or early November. Seed rate is 7–8 kg/ha for broadcast method and 4–5 kg/ha for line sowing. A spacing of 30 cm between lines and 30 cm between plants is normally adopted.

After Cultivation

Germination takes five to ten days depending upon the moisture content of the soil. Thinning is an important cultural practice to ensure uniform plant growth and better development. This is normally done when the plants are 5–6 cm high, having 3–4 leaves. Thinning is continued until the plants are about 14 to 15 cm height within a period of 3–4 weeks after sowing.

Manures and manuring: Opium poppy responds remarkably to the application of manures and fertilizers which increase both the yield and quality of opium.

Farm yard manure @ 20–30 t/ha is generally applied by broadcasting while the field is prepared for sowing. Besides, 60– 80 kg of N and 40–50 kg of P_2O_s per hectare is recommended. No potash is applied. Half of N and entire P are applied at sowing time through placement and remaining half of N placed at rosette stage.

Irrigation: A careful irrigation management schedule is essential to get a good crop of poppy. A light irrigation is given immediately after sowing followed by another light irrigation after 7 days when the seeds start germinating. Three irrigations at an interval of 12–15 days are given till pre flowering stage and then irrigation frequency is reduced at 8–10 days during flowering and capsule formation stage. Normally, 12–15 irrigations are given during the entire crop period. Any moisture stress during the stage of fruiting and latex extraction may reduce the yield considerably.

Lancing and latex collection: Opium starts flowering in 95–115 days after sowing. The petals start shedding after 3-4 days of flowering. The capsules mature after 15-20 days of flowering. Lancing of the capsules exudes maximum latex at this stage. This stage can be visually judged by the compactness and a change in the colour from greenish to light green coloured ring in the capsule. The stage is called as industrial maturity. Lancing may be done with a knife having three or four equispaced pointed ends which does not penetrate more than 1-2 mm in the capsule. Too deep or too shallow incision is not advisable. Lancing may be done early in the morning before 8.00 a.m. at two days interval in each capsule. The length of the incision should be 1/3 or less than the full length of capsule.

Harvesting and Threshing

The crop is left for drying for about 20-25 days when the last lancing on the capsules stops exudation of latex. The capsules are then picked up and the plant is removed with sickles. Harvested capsules are dried in open yard and seeds are collected by beating with a wooden rod.

The yield of raw opium varies from 50 to 60 kg/ha.

20.4 PYRETHRUM (*Chrysanthemum cinerariaefolium* Vis.)

Family: **Compositae**

Among the numerous vegetable insecticides of the world, Pyrethrum, *Chrysanthemum cinerariaefolium*, *occupies* a unique place on account of its cheapness and its wide applicability. The other important species yielding pyrethroids are **C. coccineum** and **C. marscltalki**. Pyrethrum is the much preferred synthetic insecticide in view of the following facts:

1. It is one of the safest insecticides known as it has very low mammalian toxicity and is rapidly metabolized if accidentally swallowed.
2. It disturbs insects, forcing them to move out of their hiding places.
3. It possesses instantaneous knock-down effect.
4. Being a safest insecticide, it can be used to ward off insects damaging food grains and against house hold insects.

The plant is now cultivated in many parts of the world. Kenya is the largest producer now.

The plant grows to a height of about 80 cm and is bushy and rather woody at the base. The lower leaves are alternate, petiolate and divided into lobed and dentate segments, on the flowering stems, the leaves are smaller and have only 2 or 3 segments. The solitary flower head carried on a striated stalk has 2 or 3 rows of lanceolate hairy bracts forming the involucre, a single row of cream or straw coloured ligulate florets and a disc of numerous yellow tubular florets.

About 95% of insecticide activity lies in the flowers and comes from mixture of six chemical constituents, Pyrethrin I, Pyrethrin II, Cinerin I, Cinerin II, Jasmoline I and Jasmoline II, collectively called 'Pyrethrins'. Total pyrethrin content in South Indian hills varies from 0.042 to 1.9 % but average being 1.19 %. In Kenya average pyrethrin content is 1.4 and the highest content of 2.1 % has also been recorded. The pyrethrin is distributed in different parts of the flowers but is more in achenes (93 %), while the disc florets and ray florets have only traces.

Parts of the flower	Composition of flower by weight (per cent)	Percentage of total pyrethrin
Achenes	34.2	92.4
Receptacles	11.3	3.5
Involucre scales	11.5	2.0
Disc florets	25.8	Trace
Ray florets	17.2	Trace

Climate and Soil

Pyrethrum thrives in cool dry climate of South Indian hills above 1800 m MSL and Kashmir valley. Low night temperature favours the flower production. It is sensitive to frost in its early stages. An annual rainfall of 100 cm, distributed over a period of 10 to 11 months is desirable.

Pyrethrum can be grown in a wide variety of soil but good drainage is essential. In South India it is grown in black or red loams having a pH of 5 to 5.7.

There is no named variety available, however, IIIM, Jammu has released one composite variety viz C.793 which is high yielding and having higher pyrethrin content.

Nursery: Standard nursery beds of 6.0 m length, 1.2 m width and 10 to 15 cm height are formed and mixed thoroughly with well decomposed farm yard manure. Four to five such beds may be required to produce sufficient seedlings for planting one hectare. 50 g of seeds are sown per standard bed. Pretreatement of seeds with cold water for a period of 6 to 12 hours results in good germination. After sowing, the seeds are covered with a thin layer of earth and pressed gently and a layer of grass 5 to 8 cm thick laid directly on the soil. The beds should be watered daily from a very fine rose can. Germination starts in about 8 days and is complete in about 15 to 21 days. As soon as germination is more or less complete, the grass cover should be removed. The young seedlings are highly susceptible to damping-off and if any sign of this is observed before the germination is complete, the grass cover should be removed and sun light allowed falling directly on the seedlings. The seeds are usually sown in February-March and the seedlings will be ready for transplanting within 60-75 days.

Pyrethrum can be also propagated from 'splits' obtained from older plants. The advantage in planting splits over seedlings is that they establish themselves much quicker, especially if the planting is followed by a spell of wet weather and they come to production earlier than seedlings and producing higher yields during the first year. Seedlings, however, survive much better than splits in dry weather.

Preparation of Land

The land is ploughed two or three times to a fine tilth. Sloping ground is well terraced, or graded contour trenches are formed at suitable intervals. In the Nilgiris, in addition to graded contour trenches, uncultivated grass belts of 1.5 m to 3.0 m width are left alternating with cultivated strips which are 3.2 m to 9.0 m in width to prevent any soil wash.

Transplanting: April-May planting immediately after the premonsoon showers is the best time for transplanting in South Indian hills. Seedlings with atleast 3 pairs of leaves are carefully lifted from the nursery beds and taken to the planting area in baskets. The planting holes must be dug deep enough to receive the whole root without the roots being twisted upwards. The spacing normally adopted is 60 cm between rows and 30 to 45 cm within the row. After the roots have been inserted as straight as possible, the soil must be pressed firmly around. It is advantageous to cut back newly planted pyrethrum when about three stalks have developed, because this encourages the seedling to shoot forth with great vigour. The seedlings should establish themselves under normal climatic conditions in about a month's time.

Weeding: Weeding is best done by hand pulling at regular intervals before the plants flower. Normally two weeding are done during the year of planting; the first weeding early in July and the second weeding after the South-West monsoon is over i.e. by September-October. From the second year onwards, it is necessary to do three weeding a year, the first weeding in July, the second in September-October and the third in June.

Manuring: Pyrethrum responds to phosphatic fertilizer applications. In South India, a fertilizer dose of 20 kg N, 120 kg P_2O_5 and 20 kg K_2O is normally recommended for a hectare plantation. In Kashmir, application of 60 kg N and 40 kg P_2O_5 is recommended per hectare. In Kenya, small quantities of phosphates are applied in and around the planting hole to increase flower production. Liming the soil at 100 percent of the soil requirement is ideal to improve the crop yield.

Picking

The keeping quality of pyrethrum flowers and the pyrethrin content depend entirely on the stage at which the flowers are picked. Immature flowers and over mature and full blown flowers contain less pyrethrin, and it is therefore absolutely necessary to pick the flowers at the correct stage; i.e. the flowers have atleast 3 to 4 rows of disc florets open, Harvesting interval varies from 7 to 18 days in a locality. In Kenya, the flowers are picked at intervals of 10 to 14 days, in Dalmatia once a week, and in Japan at intervals of 15 to 18 days. In South India, it is picked at 14 days interval.

The flower-heads are picked from the stalks using the thumb and index finger by a gentle jerk without causing any injury to the plants. The flowers are usually carried in baskets so as to allow aeration. If the flowers are gathered in a closed vessel and compacted, the pyrethrin may get decomposed due to the heat produced.

The flowers may be dried in the sun or in specially constructed kiln driers. Sun drying will usually take about 4 days and is possible only in dry weather and if the acreage is small. During wet weather, and if the quantity of flowers picked is large, it is necessary to have kilns for drying.

Yield

Pyrethrum plant starts flowering in about 6 to 9 months from planting in South Indian hills and also flowers throughout the year. While in Kashmir, it starts flowering only during second year of planting that to, only for 3-4 months during summer. The average yield in Kashmir is about 250 kg/ha while in South India, it varies from 180 to 400 kg/ha. Normally, pyrethrum plantation can be retained for 5 to 6 years and under South Indian hills, a period of 3 1/2-4 years is found to be economical.

20.5. SARPAGANDHA (*Rauvolfia serpentina Benth*)

Family: **Apocynaceae**

The dried root of **Rauvolfia serpentina** commonly known as serpentine root or serpentina root' is one of the most important crude drugs which has been used in the indigenous system of medicine from ancient times. The importance of this root drug and the alkaloids obtained from it has been recognised in allopathic system in the treatment of hypertension or as a sedative and tranqualising agent. A large number of alkaloids have been isolated from the roots of this plant such as ajmalicine, ajmaline, ajmalinine, rescinamine, reserpine, reserpinine, serpentine, serpentinine etc. Before 1955, the whole material was obtained from wild forest sources and thereafter attempts were made to cultivate this plant. India exports annually about 4 tonnes worth 0.43 million rupees to other countries.

R. *serpentina* is indigenous to India and its neighbouring countries. It grows wildly in the Gangetic plains, lower hills of Himalayas and also in Western Ghats.It is a perennial under shrub growing up to 0.5 m in height. It is having long elliptic lanceolate or obovate leaves occuring in whorls of three to five at the nodes of short terete stem. It bears many flowered corymbose type of inflorescence with white or pink flowers. The root system consists of a prominent tuberous soft tap root up to 6 cm in diameter when fresh and having a corky outer bark with longitudinal fissures.

Varieties

RS-1

Released by JNKVV College of Agriculture, Indore. Seeds gives good germination even after storage (for 7 months), yields air dried roots (2.5 t/ha) containing 1.64 to 2.94 % of total alkaloid at 18 month of age.

CIM- Sheel

Released by CIMAP, Lucknow is a high yielding variety.

Climate and Soil

Sarpagandha grows in a wide range of climatic conditions. However, hot and humid tropical regions with sufficient rainfall are the most suited places where it flourishes luxuriantly and can be grown in open or partial shade. A range of temperature of 10-30°C is found favourable for the plants. It grows from sea level to 1300 m elevation.

The plant grows in a variety of soils ranging from sandy alluvial loam to red lateritic loam having acidic to neutral reaction.

Propagation

(a) **Seed:** Direct sowing in the field does not give good results as the germination rate is poor (10-50 per cent). Heavy and mature seeds give high percentage of germination (43 per cent) while the light seeds which constitute 90 per cent do not germinate. The ripe seeds collected from the beginning of June to the end of October retain their viability for six months. Seeds are sown in raised beds at a depth of 1 cm in the nursery in lines 10 cm apart with 5 cm distance from seed to seed. About 5.5 kg of seeds sown in 0.05 ha area of nursery give adequate number of seedlings to plant one hectare. The seeds germinate within three weeks.

(b) **Root cuttings:** The large tap roots as well as lateral secondary rootlets are employed for preparing the cuttings of 2.5 to 3.0 cm. Planting is done in holes of 5 cm deep at the advent of monsoon and covered with 2.5 - 5.0 cm top layer of soil. The cuttings sprout within three weeks if good moisture is maintained during the period.

(c) **Stem cuttings:** Stem cuttings of 15-20 cm length with three internodes should be planted in the month of July-August in the nursery and kept moist. They strike roots within 60 days and afterwards they **are** transplanted to the main field. About 40 to 65 percent success can be expected by this method.

Planting

College of Agriculture, Indore has identified one improved selection by name RI 1. June and July are suitable months for transplanting seedlings and vegetatively propagated plants. The seedlings which are 7.5 to 12.0 cm high are carefully dug out from nursery beds and planted. The fields are irrigated soon after. Regular irrigation, weeding and manuring are required for optimum growth.

After care: The use of organic manure, leaf mould and compost are recommended @ 25-30 tonnes per hectare. Besides, a basal dose of 20 kg N, 30 kg P_2O_5 and 30 kg K_2O is applied per hectare and two top dressing of 20 Kg may be applied annually during the growing season.

Irrigation: The transplanted seedlings require irrigation at regular intervals. The crop may be irrigated fortnightly during hot dry season and once a month during winter. Under rainfed conditions yields are very low.

Intercropping: Although sole cropping of **R. serpentina** always results in more root yield, intercropping with soyabean in wet season and garlic/onion in winter is recommended. Besides **R. serpentina** can grow **well** under banana, papaya or in mango orchard. However, the plants are more robust in open than in shade.

Harvesting

However, the ideal age for the harvesting of **R. serpentina** for getting the exploitable size of the roots is after two or three years from the planting. In addition to the thick tap root, fibrous secondary roots should also be collected as they are rich in the alkaloid content. As the roots penetrate quite deep into the soil, digging forks are required to dig out the roots. Irrigation before the digging will facilitate easy picking of main as well as secondary roots. The root-bark should not be damaged during harvesting since the bark contains higher alkaloid percentage compared to the wood.

Drying and Storage of Roots: The harvested roots are thoroughly dried before storage. After air-drying, the roots are artificially dried so as to reduce the moisture content to about 3 per cent. The dried roots are then broken into small pieces of 15-20 cm and packed in air-tight containers for storage in a cool, dry place. Roots stored in godown may be periodically exposed to the air to avoid mould formation and insect damage.

Yield

From a crop of good stand about 2000 kg of dried roots can be obtained from one hectare. These roots, particularly its bark are used entirely for the extraction of the pure alkaloids. The average total alkaloid yield is 2.4 percent in the root-bark as compared to 0.40 per cent in the root wood. The fibrous roots contain an average of 2.52 percent of the total alkaloids, while stem and leaves also contain alkaloids in smaller amount ranging between 0.45-0.54 per cent. The total alkaloid content of the roots collected from various sources also varies between 0.7-3.0 per cent.

Plant Protection

R. serpentina is susceptible to many diseases especially to leaf spot, leaf blight, powdery mildews and die back diseases. 2-3 sprayings with 0.5 % Bordeaux and spraying Dithane Z.78 (0.2 %) is recommended to control the die-back. Among the pests, caterpillars are known to roll the leaves and feed on tender leaves and cause defoliation. Spraying dimethoate 0.03 % is recommended.

20.6. SENNA (*Cassia angustifolia* Vahl.)

Family: **Leguminosae**

The leaves and pods of senna (*C. angustifolia*) contain sennosides A, B, C, D, which are well known for the preparation of laxatives and purgatives all over the world. India holds leading position in the production of senna crop and export of its produce to the world market, annually earning nearly Rs 45 million. Almost all the senna leaves produced in India are exported to foreign countries and the major portion is transported to London market. The crop is grown in about 10,000 ha, mainly in Southern districts of Tamil Nadu viz Tirunelveli, Ramanathapuram and Madurai districts.

Cassia angustifolia is native of South Africa and is an erect shrub seldom reaching more than 70 cm in height. The leaves are pinnate with narrow acute lanceolate and glabrous leaflets. The flowers are brilliantly yellow in colour and borne in recemose inflorescence. The pods are flat thin and contain 5-7 dark brown seeds. Although all parts contain sennosides, the leaves and pods contain maximum content. It ranges from 1-5.3 % in Indian senna. The Alexandrian senna (*C. acutifolia*) which grows wild in Africa and Sudan contains 4-4.5 %.

Climate and Soil

Senna is a legume but produces no nodules for fixing atmospheric nitrogen. It grows well on sandy-loam with a pH ranging from 7 to 8.5. It is very sensitive to waterlogged conditions, heavy rainfall and low temperature. Once established, the crop withstands moderate saline conditions.

Varieties

ALFT2 - It is a selection and late flowering type, released from Gujarat Agricultural University, Anand. This variety remains in vegetative phase till 100 days and it is recommended for growing exclusively as a leaf crop.

Sona- It is an open pollinated variety developed by CIMAP, Lucknow and recommended for North Indian plains. The plants are120 cm tall and take 110-115 days for flowering. It gives yield of 1.1 tonnes and pod yield of 0.4 tonnes from hectare. The sennoside content in the leaf is 3.51 %.

KKM-1-A selection from the local type collected from Tenkalam area. This variety grows up to 80-100 cm.. It yields about 9118 kg/ha leaf yield and 352 kg/ha pod yield. It is suitable for stripping at an interval of 30 days in crop duration of 120-125 days. The total sennoside content is 2.54 %.

Land Preparation and Sowing

The land should be ploughed and exposed to sun for 2 to 3 weeks. The lands are so prepared to provide outlet for excess rain water at early growth period. Seed rate is 5 kg/ha. Treatment of the seeds with protective fungicides (thiram, captan or agrasan @ 2.5 g/kg of seeds to protect against damping off) and sowing at the optimum time and proper depth are the three most important factors which ensure a good and uniform plant population in the field. The seeds are sown in line at 30-40 cm apart at 1.5 to 2.0 cm depth. Light pre monsoon showers helps in germination and initial establishment. Usually two sowings are done in Tirunelveli tract one during February-March and the second one during July- November. Sowing during heavy rainy season is not advocated. Germination commences in 13 to 15 days of sowing and is over in another one week. The seed has a thick seed coat and can remain in field in hot weather without any injury but sprouted seed can seldom withstand the lack of soil moisture and usually dies. The crop is thinned at 30 days to maintain a plant to plant distance of 30 cm in the rows.

Manuring: Farmers in Tamil Nadu use 4 to 5 cartloads of well rotten farm yard manure per ha at sowing. Trials on use of inorganic fertilizers at several locations in India revealed that the crop takes away 50 to 100 kg N, 20-50 kg P_2O_5, and about 20 kg of K_2O /ha in a growing period of 4 to 5 months depending upon growth and number of pickings. In general, 80 kg of N and 40 kg of P_2O_5 may be applied to this crop. Of this, 40 kg of nitrogen and the entire dose of P_2O_5 may be given at sowing and is placed at 4-5 cm deeper below the seed so that it is easily available to the growing seedlings. The remaining quantity of 40 kg of nitrogen is given 35-40 days (just after thinning), 80-85 days, and 105-110 days age (i.e. after first and second picking of the leaf crops) in equal doses. Urea may preferably be used by broadcasting in rows and mixed thoroughly in the soil.

Irrigation and Interculture: Senna could be economically grown under rainfed conditions. An average rainfall of 25 to 40 cm, distributed from June to October is sufficient to produce good harvests. As irrigation leads to improvement in yield, wherever easily available, it is given at 40 days, 75 days and 100 days age when plants bear new growth of foliage and flowers.

Senna is hardy and requires first weeding-cum-hoeing at 25-30 days, a second at 75-80 days and a third at 110 days from sowing.

Harvesting

Young senna leaves and pods contain a high sennoside content but since the produce is sold on the basis of weight, a balance between weight and content is to be made to choose its stage for harvest. First picking starts at 50 to 70 days age, depending upon total plant growth. A second picking be taken at 90 to 100 days and the third picking between 130 to 150 days when the entire plants are removed so that the harvested material includes both leaves and pods together. Although, root- bark contains sennoside, it has not yet come as an article of trade.

The harvested crop should be spread in a thin layer in open area to reduce the moisture. Further drying of the produce is done in a well-ventilated drying sheds. It takes 3 to 5 days to dry the produce in the sheds. The dried produce usually possesses 8 per cent moisture. The properly dried leaves and pods should have light-green to greenish-yellow colour. Improper and delayed drying changes the colour to black or brown which lowers the sennoside content and thus the price. Seeds contain no sennosides and it adds weight to the produce only.

Yield

A good average crop of senna can give 1500 kg/ha of dry leaves and 700 kg/ha of pods under irrigated and good management conditions. The yield under rainfed conditions is about 1000 kg of leaves and 400 kg of pods.

Plant Protection

Dry-rot kills the plants. The fields are drenched with any organo fungicides but it gives only a partial control. Leaf spot and leaf blight are the other diseases, caused by *Cercospora spp.* and *Phyllosticta spp.* respectively, usually occurring at later stage of growth, when atmospheric humidity is high. Spraying of 0.15 per cent Diathane M.45 at fortnightly interval for 3 times in a period of 5 to 6 weeks is recommended to control these leaf diseases.

20.7. ISABGOL (*Plantago ovata* Forsk.)

Family: **Plantaginaceae**

Isubgol or Psyllium (*Plantago ovata*) is important for its seed and husk which have been used in the indigenous medicine for many countries. It has the property of absorbing and retaining water (40-90 %) and therefore it works as an anti-diarrhea drug. It is beneficial in chronic dysenteries of amoebic and basillary origin. The seed has also cooling and demulcent effect and is used in ayurvedic, unani and allopathic medicines. The husk yields a colloidal mucilage consisting mainly of xylose, arabinose and galacturonic acid. In India it is grown in about 16,000 to 20,000 hectare in North Gujarat and it is also recently cultivated in small areas in Rajasthan, Haryana and Bihar. The husk and peel are exported largely to USA, West Germany, the UK and France, fetching a foreign exchange of more than 10 crores annually.

It is a stemless annual herb often attaining a height of 30 to 40 cm, with rosette leaves. The plant bears erect ovoid or cylindrical spike with minute white flowers about 45-68, protogynous. Fruit is capsule, each seed is encased in a thin, white, translucent membrane, the husk, which is odourless and tasteless.

Climate and Soil

It requires cool and dry weather and hence in India, the crop is grown in winter i.e. from November - December to March - April. Humid weather at maturity results in shattering of seeds. A light well drained sandy loam to rich loamy soil with a pH of 7 - 8 is ideal.

Varieties

Variety	Breeding Method	Parents	Important traits	Institute/ University
Gujarat Isabgol-1 (GI-1)	Introduction	Selection from introduced material	Dark green leaf, moderate tillers and medium spike length (4.0-4.5 cm). Height about 50 cm, matures in 110-115 days	AICRP on M &AP, Center, Gujarat Agricultural University, Anand
Gujarat Isabgol-2 (GI-2)	Mutation	Mutant from GI- 1	Medium broad and pale green leaf, medium long spike (3.8-4.6 cm), Height 58-64 cm, matures in 110-115 days.	
Haryana Isabgol-5 (HI-5)	Selection from GI 2	GI-2	Maturity: 140-145 days Yield 1000-1200 kg ha^{-1}. Husk: 25-30 %	AICRP On M&AP, CCS Haryana Agricultural University, Hisar
Jawahar Isabgol-4 (JI-4)	Selection from germplasm	Selection from local bulk	Leaf: Narrow, dense lathery, dark green and hairy. Ovary: Violet pink Seed: boat shaped, hard, light pink	AICRP on M&AP, Mandsaur

			Seed husk: Rosy white membrane like. Yields 1300-1500 kg/ha.	
Gujarat Isabgol-3	Selection	Selection from local germplasm collections	Long, thin and dark green leaf, profuse tillers and long spike (4.5-5.1 cm). Height 54-61 cm, matures in 106-112 days and gives average yield of 1284 kg ha^{-1}.	S.K. Nagar Krishi Viswavidyalaya, Gujarat
Niharika	–	–	A high yielding variety. 120 days maturity, grows as winter crop for North Indian plains.	CIMAP, Lucknow

Preparation of Land

Field must be free of weeds and clods and should have fine tilth for good germination. The land is laid into flat beds of convenient sizes i.e. 1.0 m × 3.0 m or 2.5 m or 2.5 m.

Sowing: Fresh seeds from the preceding crop season should be sown for getting high percent germination. The seed rate varies from 4 - 6 kg and is sown after pretreatment with thiram @ 3 g/kg of seed to protect the seedlings from the possible damage of damping off. The seeds, being small and light are mixed with sufficient quantity of fine sand before sowing. The seeds are sown broadcast and are swept lightly with a broom in one direction to cover them with some soil.

After Cultivation

Timely weeding is important to encourage good growth of the plants. After 20-25 days of sowing, first weeding is done and 2-3 weeding are required within 2 months of sowing.

It responds to manuring. 25 kg N/ha and 25 kg P/ha are applied as basal dose at the last ploughing and another dose of 25 kg N/ha is top dressed 30 days after sowing.

Immediately after sowing light irrigation is essential. First irrigation should be given with light flow of water. The seeds normally germinate in

6-7 days. If the germination is poor, second irrigation may be given. Later on, irrigations are given as and when necessary. Last irrigation should be given at the time when maximum number of spikes has reached the milk stage.

Harvesting and Processing

The crop will be ready in about 110-130 days after sowing. When mature, the crop turns yellowish and the spike turns brownish. The seeds are shed when the spikes are pressed even slightly. At the time of harvest, the atmosphere must be dry and there should not be any moisture on the plant.

The plants are normally cut at the ground level or are uprooted if the soil is loose textured. The harvested plants are threshed and winnowed, and the seeds repeatedly sifted until clean. The seeds may be marketed whole or the husk may be sold separately. Seeds are fed to a series of shellers, in each sheller the grinding pressure is so adjusted to remove only the husk. This is separated by fans and sieves at each sheller and the ungrounded material is sent to the next sheller. The husk: seed ratio is 25:75 by weight.

The average yield is about one tonne of seeds per hectare.

Plant Protection

Downy mildew caused by *Peronospora plantaginia* is the serious disease at the time of spike initiation. Spraying of copper oxy chloride or Dithane Z.78 @ 2.0 g/L of water is recommended as a prophylactive measure on 30th day from sowing and repeated twice at an interval of 15-20 days.

20.8. MEDICINAL SOLANUM
Solanum khasianum (Clarke)

Family: **Solanaceae**

Solanum khasianum (Syn. *Solanum viarum)* serves as a supplementary source of steroidal raw material for industries in India. Solasodine obtained from its berries are used as a substitute for diosgenin in the synthesis of steroidal hormones. It is used in the manufacture of oral contraceptive tablets. The extract also possesses some nematicidal and bactericidal properties.

The genus *Solanum* comprises of about 2000 species which can be broadly grouped as tuberous group and non-tuberous group. *S. viarum* belongs to the latter group and is native of India and occurs in nature in Sikkim, West Bengal, and Orissa and in Western Ghats up to 1600 m MSL.

It is a stout, much branched undershrub varying in height from 0.75 to 1.40 m with more of straight prikles often mixed with a few curved spines on the stem. Leaves are ovate, lobed, hirsute and prickly on both the surface. The flowers are white, borne in racemes with 1-4 flowers and the berries are yellow or green in colour, globose in shape. The solasodine content in the berries varies from 1.00 to 1.75 %.

Besides, *S. khasianum,* many other species of *Solanum,* indigenous to India, also contain solasodine in their fruits

	Species	Solasodine content (%)
1.	S. *mammosum*	0.80–1.00
2.	S. *pubescence*	0.74–0.95
3.	S. *saeforthianum*	0.75–2.00
.4.	S. *eleaegnifolium*	1.50–1.70
5.	S. *incanum*	1.80–2.00

Varieties

IIIM, Jammu has identified two mutants which are high yielding with high solasodine content.

1. **RRL-20-2:** It is vigorous in growth with 3-4 fruits per node, a yield potential of 6-7 tonnes of fruits and 42-45 kg of solasodine/ha.
2. **RRL-GL-6:** It is a spineless mutant. It is almost similar to RRL-20-2 in all aspects except having slightly reduced solasodine content.
3. **Arka Sanjeevani:** An inter varietal hybrid between Galaxo mutant X BARC mutant developed by IIHR, Bangalore. It is the least spiny hybrid with curved spines suitable for high density planting. It gives threefold increase in berry yield and solasodine content.
4. **Arka Mahima:** An induced tetraploid recording higher solasodine content (2.88 %) than diploid counterpart Arka Sanjeevani (1.99 %) and is devoid of spines. It is released from IIHR, Bangalore.
5. **Galoxo Mutant:** A less spiny mutant evolved by mutation breeding using wild solanum as base material. This mutant is characterized by presence of well-developed straight spines on the laminar surface while the stem is devoid of spines. Yield of fresh berries is 11.92 t/ ha.
6. **BARC-1 Mutant:** A curved spine mutant developed through irradiation of dry seeds with 10 Kr gamma rays. It has high berry yield and glycoalkaloid content.

7. **IIHR 2n-11:** Completely devoid of spines, produce high berry yield at high density planting containing 2.5 to 3 % of solasodine.

8. In *Solanum laciniatum,* an exotic selection (NH 88-12) with higher dry berry yield (31.38 q/ha) and solasodine content 4.18 % (in berries) and 1.74 % (in leaves) has been identified.

Seeds and Sowing

S. khasianum is either directly sown or transplanted. For transplanted crop, seeds are sown in seed bed in February- March. The seeds are sown in lines 10 cm apart and covered with a thin layer of soil. The seed beds should always be kept moist by sprinkling water as and when necessary. Scarification of seed with 5 % nitric acid improves the germination of the seeds. About 1.25 kg seeds provide enough seedlings for planting one ha of land. Germination is complete in 7-10 days. 4-5 weeks old seedlings having about six leaves are ready for transplanting.

Seedlings may be transplanted in moist soil at spacing in 50 × 50 cm. For intercropping and easy cultural operations, a wider spacing of 90 × 150 cm is ideal.

After Cultivation

S.khasianum responds to the application of manures and fertilizers. Nitrogen, phosphorus and potash at the rate of 100, 60 and 40 kg/ha is recommended. While the entire quantity of P and K are applied at the time of field preparation, N is generally applied in three splits at different intervals.

Irrigation is generally done initially and subsequently the crop is grown as rainfed. In general, a moisture stress inhibits vegetative development, fruit production and yield but the solasodine content increases with moisture stress.

Harvesting and Processing

Harvesting is one of the difficult operations in the cultivation of *S. khasianum.* The spiny nature of the plant hampers plucking of fruits and hence the use of gloves is helpful. Plucking berries at the right stage of maturity is very important. The solasodine content increases consistently from the early stages of fruit development and attains a maximum when the fruit colour changes from green to yellow and it is considered as the right stage for harvesting the berries. Berries contain 70- 75 % solasodine and 60 % of this is present in the seeds and the rest 40 % in the pericarp.

Water content in the berries is about 70-75 % percent and hence it should be dried immediately. As slow drying causes a loss in solasodine due to degradation, oven drying is suggested. To hasten drying and to impart better colour, fruits are cut into halves and spread in thin layers on the floor and turned frequently. When the dried berries make cracking sound they are packed in bags. Dried berries should posses not more than 10-12 percent moisture and not less than 2 percent solasodine.

On an average, 6-8 tonnes of berries/ha can be obtained.

20.9. PERIWINKLE (*Catharanthus roseus* G. Don) (*Syn. Vinca rosea*)

Family: **Apocyanaceae**

Periwinkle is a perennial ornamental herb found throughout India on waste lands and sandy tracts. It has medicinal importance owing to the presence of indole alkaloids raubasin (ajmalicine) and serpentine in its root which has anti-fibrillic and hypertensive properties. The leaves contain two alkaloids viz vinblastine and vincristine which form the constituents of patented cancer drugs and vincristine alkaloids are distributed in different parts of the plant but the roots contain the maximum (0.75-1.20 %), followed by the leaf (0.60-0.65 %).

USA imports about 1000 tonnes of leaves while West Germany, Italy, Netherlands and UK import about 1000 tonnes of roots. There is ample scope for this crop if the raw meterials are used in the manufacture of drug in India itself as there is a decline in the demand for the raw materials outside India. Farmers may also prefer this crop because of its wide adaptability, ability to grow in marginal lands and drought hardiness.

It is a perennial herb, often grows in garden for its pink and white flowers which bloom throughout the year. It bears flexible long branches with simple opposite leaves. Flowers 2-3 in cymes, axillary and terminal clusters. Fruit is a cylindrical follicle with many black seeds.

Varieties

There are no recognized varieties but there are three local types based on the colour of the flowers viz *alba* with white flowers, *roseus* - with pink rose coloured flowers and *ocillata* with white flowers having rose-purple spot in the centre are recognized.

Nirmal- White flowered variety developed from CIMAP, Lucknow. It yields about 1200 kg of dried leaves and 80 kg of roots/ha, resistant to collar rot and root rot.

Dhawal-A mutant bred by chemical mutagen treatment of the seeds of variety Nirmal developed by CIMAP, Lucknow. It consists of light green to greyish pubescent leaves with distinctly undulating leaf margin, green stem, white flower, leaf yield of 1352 to 2557 kg/ha, alkaloid yield 0.89 to 1.40 % (in leaves), 1.60 to 2.22 % (in roots), and resistant to die-back disease.

Prabhat- It is developed at HAU, Haryana. Duration is 8-10 months, dry root yield is 15-18 q/ ha, dry leaves yield is 20-25 q/ha. It is dark purple stemmed, shining leaves and pods with dark pink flowers.

Climate and Soil

The cosmopolitan distribution of the plant shows that it has no specific climatic requirements. Its natural environments are, however, tropical and sub-tropical areas. A well distributed rainfall of 100 cm or more is considered ideal for raising it as a commercial crop under rainfed conditions.

Similarly, it grows on any type of soil except those which are highly alkaline or water-logged. It grows wild in coastal area. Light sandy soils rich in humus are preferred for large scale cultivation of the plant.

Propagation

The plant is propagated from seeds. Fresh seeds are preferable as they lose viability on long storage. Seeds can be sown directly in the field or the plants raised in the nursery and transplanted later on.

Direct sowing is to be done for plantations of a large area, as it reduces the cost of sowing. About 2 to 3 kg seed are required for raising one hectare. The seeds are mixed with sand about 10 times its weight for even distribution and are sown during beginning of monsoon in rows 45 cm apart. When the plants grow up they are thinned out, leaving a distance of 25-30 cm between the plants.

For nursery sowing and transplanting, about 500 g of seed sown in 200 m^2 bed is required for producing seedlings for one hectare. The seeds are sown in well prepared beds during March/April in rows about 1.5 cm deep, covered with light soil and leaf mould mixture and are watered to keep the bed moist. In about 10 days time the seeds germinate and in 2 months time (height 6-7 cm) they become ready for transplanting. In the field, the seedlings are transplanted at a spacing of 45 cm × 30 cm or 45 cm × 45 cm.

After Cultivation

The crop requires two weeding, the first one about 60 days after sowing/transplanting and the second one in another 60 days.

The plants do not require much water as they have drought resistant capacity, in areas, where rainfall is evenly distributed throughout the year, no irrigation is required, but in areas where monsoon is restricted, 4 to 5 irrigations are required during the life of the plant to get good yield.

They are not generally manured, however, for getting a good yield of both leaves and roots, farmyard manure at about 15 tonnes per ha should be applied and a fertilizer mixture of N (50 kg), P_2O_5 (75 kg) and K_2O (75 kg) per hectare is applied as a basal dose.

Harvesting

The crop becomes ready for harvest of roots after one year. But two leaf stripping can be taken, the first one after 6 months and the second after 9 months of sowing. Third stripping of leaves can also be taken when the whole plant is harvested after one year. For seed collection, matured fruits are hand-picked and dried in shade and threshed lightly. This method ensures mature seeds with even germination. But, the usual practice is to uproot the plants, dry them in shade and thereafter thresh lightly for seeds. The seeds obtained by this method are not uniform and their germination is poor.

For harvesting of roots, the crop is cut about 7.5 cm above the ground and dried for stems, leaves and seeds and then the whole field is copiously irrigated and ploughed and the roots are collected. The roots are washed well and dried in shade and later made into bundles for marketing.

Yield

Under rainfed conditions about 0.75 tonne of roots, 1.0 tonne of stems and 2 tonnes of leaves (all dry basis) may be obtained from one hectare. But under irrigated conditions, 1.5 tonnes each of roots and stems and 3 tonnes of leaves per ha can be obtained.

20.10. BELLADONNA (*Atropa belladonna* L.)

Family: **Solanaceae**

Belladonna is an important source for tropane alkaloids which include hyoscine, hyoscyamine and atropine which are used in medicine because of their antipholinergic and roots of Indian belladonna *(A. acuminata)* and European belladonna *(A. belladonna)* are used in the pharmaceutical industry. The Indian belladonna is found wild in the forest of Western Himalayas while European belladonna is grown in Italy, Yugoslavia, USSR, Rumania and United Kingdom. These species are tall herbs bearing yellow *(A. acuminata)* or purple flowers *(A. belladonna).* India is importing about 13 tonnes of belladonna extract worth of nearly 37 lakhs to meet our domestic requirement as the supply of belladonna from natural source is undependable and not at all sufficent.

It is a temperate cool season crop found growing successfully above 1500 m MSL in Northern hills. Berries are collected and crushed gently to separate the seeds and dry them. Good healthy and heavy seeds are sown during June-July in raised beds at the rate of 1 kg per 100 m^2. The seeds take 4-6 weeks to germinate and the seedlings when they attain to a height of 15-20 cm are transplanted in the field at spacing of 60 × 60 cm or 60 × 45 cm and irrigated till they establish. A basal dose of 2-3 tonnes of FYM besides 60 and 40 kg of N and P/ha are applied. Again 30 kg N/ha is given in 4-5 splits as top dressing after every harvest.

The alkaloid content increases steadily from transplanting till flowering stage. The harvesting consists of cutting the leaves about 30 cm above ground level using pruning scissors. A total of 5-6 cuttings can be obtained normally. The leaves so harvested are sun-dried by spreading them in thin layers and turning them frequently. The woody stems are discarded before drying. On an average, 200 to 400 kg of dried leaves can be obtained from one hectare.

20.11. RYE ERGOT

Ergot refers to the slender blackish to brown coloured sclerotial body of the fungus *Claviceps purpurea.* It develops on rye spikes or any wild grass infected by the above fungus. The alkaloids obtained from the ergot are used in obstetrics making child birth easy and stoppage of bleeding after the child birth. It is also used against migraine headache and lowering hypertensions.

Selection of Host Plant

Ergot fungus is known to infect many wild and cultivated grasses but the best for the large scale production of ergot in terms of quantity and quality is the rye *(Secale cereale).*

Rye is a cool season crop and in South India it can be grown in hill stations above 1500 m MSL. It even comes up during dry winter months. It is a hardy crop and is not exacting in its soil requirement however, loamy soil is best suited for its growth.

The land is prepared thoroughly to make fine tilth and is sown in furrows of 4-5 cm depth and 25 cm apart. 70 to 80 kg seeds are required for one hectare. Best sowing time is September to October in Kashmir valley and November in South Indian hills. The rye crop is given a fertilizer dose of 16 kg N, 3 to 4 kg P and 20 kg of K_2O per hectare. Half the quantity N, full dose P and K are applied as basal and the remaining half of N is applied as top dressing just 7 to 10 days before flowering.

Inoculation of the Rye Crop

Rye spikes are to be inoculated two times with the culture of the Ergot fungi. The stage of inoculation and method of infection of the rye crop is very critical for the large scale production of rye ergot. For inoculation, 40-50 bottles (500 ml capacity) of 40-50 days old inoculum on sterile wheat grains are mixed in 125- 150 litres of clean water. This is sufficient to inoculate one hectare rye crop, using needle board puncture method. Field scale inoculation of rye spike is achieved by dipping the sponge board in a container having the ready inoculam and is held in the hand against a group of rye spikes and punching of the spikes against the sponge board is done with the help of needle board in the other hand. The above operation may be repeated as frequent as possible. The correct stage for the first inoculation of the rye spikes is when more than 20-30 % of the rye spikes have 1/3 rd to 1/2 portion of the spikes emerged out of the boot leaf and leaving rest of the basal portion still lodged within the boot leaf. The second inoculation may be repeated 10-15 days after the first inoculation when 70-80 % of rye spikes have appeared

Collection of Ergot Sclerotia

Collection of ergot sclerotia is generally done 8-10 weeks after the inoculation or 15 days before the ripening of rye grains. Well developed sclerotia may be picked 15-20 days earlier to the regular collection. Harvesting is generally done by the hand picking. If successful inoculation was there, about 4 to 10 kg of sclerotia may be collected from every 100 kg of rye grain which works out to 100 to 200 kg/ha. Besides, 400 to 500 kg of rye grain may also be harvested.

Proper drying of such harvested sclerotia is essential to have pharmaceutical quality grade. If weather is not very sunny, the sclerotia is spread thinly on a cement surface or on a canvas with frequent stirring up to 4-6 days. If there is a cloudy weather, drying is done in well ventilated rooms where it may take 10-15 days. Well dried ergot is packed in medium sized metallic or wooden container having water proof alkathane inner-liner.

20.12 COLEUS (*Coleus forskohlii*)

Family : Lamiacaea

Coleus forskohlii Syn. *Coleus barbatus, Plectranthus forskohlii*, a native of India, is a well known plant through out the country and one of the most significant medicinal crops for its tuberous roots. The dried roots are found to be a rich source of forskolin and are used for treating hypertension, glaucoma, asthma, congestive heart failures and certain types of cancer. The tuberous roots, resembling a carrot in shape and brown in colour, are the commercial parts. The plant is known as 'Pashanbhedi' in Sanskrit and 'Patharchur' in Hindi. Recent discoveries have indicated that the forskolin is useful against cholesterol and also used in cosmetics. The species came into commercial cultivation after the discovery of forskolin, a unique adenylate cyclase activating drug which is highly useful in activating the cardio vascular system. The dry roots contain forskolin ranging from 0.10 to 0.80 per cent. One of the Indian medicinal plants which were very little known until a few years ago has attained international importance.

There are about 10,000 ha area under this crop in the country, cultivated in parts of Rajasthan, Maharastra, Karnataka and Tamil Nadu. The annual estimated production is 2000 tonnes of dry roots.

Botany

The plants produce thick roots in the form of elongated tubers. Radially spread roots are fasciculate, succulent, tortuous with 1.0 to 3.0 cm thickness and 20 cm length. The inner roots are orange coloured and have the characteristic pungent odour. The plants have square stems branched where nodes are often hairy. Leaves are pubescent, narrowed into petiole. Though it is a biennial, it is cultivated as an annual.

Variety

CO (Col)-1

A selection from local type (Theni) released from TNAU with a crop duration of 160-180 days and tuber yield of 1.98 t/ha. It has high forskolin content (0.54%) tolerant to root rot, wilt and nematodes.

K-8

It is released from IIHR, Bangalore and a selection from Karnataka. It contain 0.5% forskolin and high tuber yield

Aisiri

A new variety with forskolin content of 0.7 % has been released from UAS, Bengaluru.

Climate and Soil

Coleus is a crop of the tropics and is found growing well in foot hills and plains under tropical and sub tropical conditions. It prefers a warm temperature (25-32°C) with moderate humidity ranging from 50-60 %. It thrives better in well drained soils with a pH ranging from 5.5–7.0. It does not require very fertile soils and can be economically grown under marginal soils.

Propagation and Planting

Coleus is propagated by terminal cuttings. Normally, 10-12 cm long cuttings comprising 3-4 pairs of leaves are preferred. These cuttings are either rooted in nursery and then planted in the main field or planted directly in the main field. The ideal season for planting is from June to July with the onset of South west monsoon. Before planting, the field is ploughed deep soon after the receipt of pre monsoon showers and brought to fine tilth. The crop loves high amount of organic manure and about 25 t FYM/ha is applied. Ridges and furrows are prepared at 60 cm spacing. The height of the ridge should be 15 cm from ground level. The cuttings are planted at 30 or 45 cm distance depending on the soil type. While planting, care should be taken to see that minimum of two nodes should be underneath the soil. Watering should be done before and after planting. Under drip irrigation system, raised broad row ridges of 90cm width are prepared at 60 cm interval and planting at 60 cm spacing between rows are planted. The space between two plants should be 45 cm.

After Cultivation

The crop requires plenty of organic manure. In addition to 25 tonnes of FYM, addition of 1 ton of vermicompost, 150 kg of neem cake, 500 kg of gypsum are applied to condition the soil and to improve its fertility by organic means. Many farmers adopt organic farming by avoiding chemical fertilizers and pesticides. 'Panchagavya' 3 % organic spray is given along with root drenching. A fertilizer dose of 40 kg N, 60 kg P_2O_5 and 50 kg K_2O per hectare is recommended for Tamil Nadu. Half the dose of N, the whole P and K may be applied as the basal dose followed by the remaining half N, 30 days after planting as top dressing.

The first irrigation is given immediately after transplanting. In the initial phase, the crop is irrigated once in three days and thereafter, weekly irrigation is enough to obtain good growth and yield.

Two or three weeding are given and after the second weeding, earthing up is given. As the roots are shallow, deep digging should be avoided.

Plant Protection

The leaf eating caterpillars, mealy bugs and root knot nematodes are the important pests. The insect pests can be controlled by spraying the plants and drenching their roots with 0.1 % chlorpyriphos, while nematodes can be controlled by application of carbofuran granules @ 20 kg/ha. Soil application of FYM @ 12.5 t/ha + 500 kg neem cake/ha + *Trichoderma viride* @ 2.5 kg/ha before planting is effective for bio management of nematode fungal disease complex involving *Meloidogyne incognita* and *Macrophomina phaseolina*.

Bacterial wilt is the major disease and can be controlled by spraying and drenching the soil with 0.2 % captan solution immediately after the appearance of the disease and later after a week's interval. Procuring planting materials from infected areas may be avoided.

Harvesting, Processing and Yield

The crop is ready for harvest in 180 days after planting. Flowers if any should be nipped-off during the growing period to obtain more root biomass. The roots are harvested either by ploughing using a bullock or by tractor. The tubers can also be manually dug and taken with least damage. The roots are cleaned making free of soil and transported for drying. The roots are cut into small pieces using mechanized rotary motors. The root bits are spread thinly on the cement yard and allowed to dry for 3-5 days. The roots get completely dried and are packed.

On an average, a yield of 1500 kg of dried tubers per hectare is obtained. If proper cultivation practices are followed, a yield of 2500 kg of dried tubers can be expected per hectare.

20.13. GLORY LILY (*Gloriosa superba* LINN.)

Family: **Colchicaceae.**

It is a native of tropical Asia and Africa. The genus Gloriosa is comprised
of about 10 to 15 known species and among them *G.superba* and *G.
rothschildiana are* found in India. Known in Kannada as 'Agnishike','
Indrana huvu', as 'Kalihari'in Hindi and in Tamil as 'Kanvazhipoo',
'Kanvazhikizhangu' has been used in the Indian system of medicine since
time immemorial. Its tubers are reported to have been used as a tonic,
antiperiodic, antihelminthic and also against snake bites and scorpion stings.
The drug is a gastro intestinal irritant and may cause vomiting and purging.
It is sometimes used for promoting labour pains and conversely also an
abortifacient. The medicinal properties of the drug are due to the presence
of alkaloids, chiefly colchicine and gloriosine. Colchicine is used in the
treatment of gout; a common disorder. The colchicine content varies from
0.15 to 0.3 % in the tubers, and in the seeds it ranges from 0.7 to 0.9 %.
The crop is grown mainly for its seeds which are in great demand within
the country and in the international market.Tamil Nadu leads in production
of glory lily in India with an estimated area of 3000 hectares with annual
production of 1000 ton dry seed.

Botany

It is herbaceous climbing perennial, growing between 3.5 to 6m in length,
but usually trained at 1.5m above ground level. The vines are long, weak-
stemmed with tuberous roots that support themselves by means of cirrhosed
tips. The leaves are ovate, lanceolate, acuminate, the tips spirally twisted to
serve as tendrils. The flowers are large, solitary. In the bud stage, the petals
hang down over the ovary and on maturity; they assume an erect position,
leaving the ovary with its stigma exposed at right angles.

Climate and Soil

It is a tropical plant and comes up well in warm dry regions receiving an
annual rainfall of 750 mm or even less is preferred, places enjoying more
humidity (75 to 80 %) are more prone to Curvularia blight causing total
loss of vine.

It prefers sandy loam soils on the acidic pH with good drainage, for
its successful growth. A soil pH range from 6-7 has been found suitable
for raising the crop.

Propagation

It is commercially propagated from its underground, V-shaped rhizomes or sexually propagated by seeds. The plants raised from seeds take nearly three to four years to flower. Hence, except for experimental purposes, seed propagation is not favoured by the growers. Glory lily produces a bi-forked tuber during the growing season and each of these forks has only one growing bud. Tubers should be handled carefully, as they are brittle and liable to break easily. If the growing bud is subjected to any kind of damage, the tuber will fail to sprout. Since the vigour of the vine and its flowering and fruiting ability depends on the size of the tubers, it should not weigh less than 50-60g. The plants raised from smaller tubers do not produce flowers during the first year. The dormant tubers start sprouting from the month of May. About 2.5 to 3.0 MT of tubers are required for planting one hectare. In order to avoid rotting of the tubers before sprouting, only healthy tubers should be selected for planting. The selected tuber pieces should be treated with suitable fungicides, preferably emisan @ 0.08 %.

Field Preparation and Planting

The field should be ploughed and harrowed several times until it is brought to fine tilth. All the grass stubbles and roots should be removed. The field must be leveled properly and drainage arrangements made to avoid water logging during the rains. The field is then divided into subplots of convenient sizes. About 15-20 t/ha of FYM or compost should be mixed well into the soil. About 30cm deep furrows are opened at a spacing of 45-60cm. The treated tubers are planted at a depth of 6-8cm, keeping a plant to plant distance of 30 to 45 cm, depending upon the type of soil. Closer spacing has been reported to favour cross pollination, thereby improving the fruit set.

After Cultivation

A fertilizer dose of 120 kg N, 50 kg P_2O_5 and 75 kg K_2O/ha is required for a good crop. Of the nutrients, the whole P_2O_5 and K_2O and one third of N is applied as a basal dose and the remaining two-third of N should be given in the first six to eight weeks after planting.

Frequent irrigation is required during the sprouting time to keep the surface soft, so that there is no hard-pan formation in order to facilitate easy sprouting and emergence of the growing tip outside the soil. Irrigation should be withheld until the flowering is over, to prevent rotting of the tubers. Excess watering is harmful to the plants and causes yellow or brown coloured patches on the leaves which fall off prematurely. Application of 150:100:300 kg NPK/ha through water soluble fertilizers is recommended for doubling the seed yield.

The provision of support is necessary for successfully growing glory lily. Since the stem is very slender, when the plants are about 30-40 cm tall, they should be staked or tied to wires or allowed to climb on some sort of frame. Various standards are used by farmers. The GI trellis wire support system is the commonly adopted practice.

The flowers have deflexed stigma which requires assisted pollination. Hand pollination is done between 7-11 am every day. Pollens are collected using brush and dusted on the just opened flowers to ensure maximum seed set.

Plant Protection

Leaf blight caused by the fungus *Curvularia lunata* is a serious one and can be controlled by spraying 0.2 % mancozeb. **Tuber rot or basal stem rotting** and **wilting** can be controlled by drenching the soil with carbendazim (0.2 %) or *Psuedomonas fluorescens* @ 0.5 %.

Harvesting and Processing

Glory lily is a crop of 180 days duration. When planted in June, it starts bearing flowers after 55 days and continues to flower and fruit till October. The fruit requires about 105-110 days from the set to reach maturity. The right stage of harvest is when the capsule starts turning light green from dark green and the skin of the fruit shows a shrunken appearance and becomes light in weight. At this stage when pressed the pod gives a crinkling sound. After picking, the capsules should be kept in the shade for 7 to 10 days to facilitate the capsule to open up displaying deep orange yellow coloured seeds. The seeds and pericarp are separated manually and dried for a week in the shade, by spreading them over any clean dry floor or any platform specially erected for the purpose. At the later stages, the seeds are dried under sun. The dried seeds are then packed in moisture- proof containers and stored until they are marketed.

The yield of seeds differs greatly, depending upon the size of the tubers used for planting and age of the tubers. The average yield is 500 kg dry seed/ha. The seed yield gradually decreases in the third year and thereof. Under drip and fertigation system, the seed yield is very high ie 1000 kg/ha.

CHAPTER

21

Other Medicinal Plants

S. No.	Common name	Botanical name and Family	Brief description and uses
1.	Achillea	*Achillea millefolium* (L) Compositae	A herb, found grown in Western Himalayas, contains a glycoside, *achillein* useful as stimulant, tonic, diaphoretic etc.
2.	Sweet Flag **T**: Vasambu **H**: Bach	*Acorus calamus* Sinn, Araceae	A semi aquatic perennial herb with creeping and much branched aromatic rhizome and fibrous root, occurs wild or cultivated in marshy places all over India, especially in the hilly tracts; economic part is the rhizome, contains essential oil and alkaloids used as a stimulant, emetic, nauseant, stomachic, expectorant, carminative etc.
3.	Vilvam	*Aegle marmelos* Rutaceae	A tropical and subtropical tree, parts used are root bark, unripe or half ripe fruit, pulp of ripe fruits, used against intermittent fevers, astringent, diuretic.

4.	Adhatoda **T**: Adadodai **H**: Adulasa	**Adhatoda vasica** Nees. Acanthaceae	A shrub, found throughout India, leaves contain an alkaloid, vasicine. Leaves and roots are used in cough, chronic bronchitis, asthma etc. stomach ache, and diarrhea, cooling and laxative.
5.	Agave **T**: Kattalai	**Agave americana** (L) Amaryllidacaeae	An exotic (America) short stemmed semi woody plant, bearing a rosette of long erect pointed and fleshy leaves, juice of leaves used as laxative, diuretic, diaphoretic.
6	**T**: Sortu Kattalai **H**: Gheekanwar	**Aloe vera** Liliaceae	A perennial , succulent plant with stout stems, the gel obtained from the stem is a potential laxative, used in minor wounds, burns and frost bite etc.,
7	**T**: Nela vembu **H**: Kalmegh	**Andrographis paniculata** Acanthaceae	An annual herb, found in dense forests, the plant is bitter, acrid, cooling, laxative, vulnerary, antipyretic, antiperiodic, anti-inflammatory, expectorant, antityphoid etc.
8	Sour sop Or Graveola	**Annona muricata** Annonaceae	An exotic (Caribbean and Central America), an evergreen tree grows to 4 m high, pharmaceutical activities include anticarcinogenic and genotoxic effect; the leaves, fruit, bark and seeds are known to contain more than 50 mono-THF acetogenins.
9.	*Ammi majus*	**Ammi majus (L)** Umbelliferae	An exotic (Egypt) biennial glabrous herb requiring mild cool climate; fruit is the economic part, containing 0.4 % Xanthotoxin used in the treatment of leucoderma and in the formulation of sun tan lotion.
10.	Shadavari	**Asparagus racemosus (Willd.)** *Liliaceae*	Roots are used- refrigerant, diarrhea, aphrodisiac and antiseptic.

11.	Neem T: Vembu H:Nim	*Azadirachta indica* **(A. Juss)** Meliaceae	A tall evergreen tree, distributed throughout India, all parts of the plants have medicinal properties, used against leprosy, intestinal worms, piles etc, seed contains 0.15 % Azadirachtin, a highly oxygeneted terpenoid which has antifeedant activity.
12.	H:Brahmi T:Neerbra mi	*Bacopa monnieri* *Scrophulariaceae*	A common creeping annual in humid and warmer parts, whole plant is used in ISM as a nerve tonic, and for epilepsy and insanity, also as diuretic, for treating rheumatism, asthma, also well for enhancing memory and vitality.
13.	Sapan wood, T:Parthangi H: Bakam	*Caesalpinia sappan* Caesalpiniaceae	A multipurpose tree, yielding valuable natural dyes with medicinal properties- the decoction of the wood is a powerful emmenagogue, because of its tannic and gallic acids, is an astringent used in mild cases of dysentery and diarrhea.
14.	Indian penny Wort T: Vallarai H : Brahama - menduki	*Centella asiatica* Umbelliferae	A common trailing herb, with kidney shaped leaves, found throughout India in marshy places up to 1600m., leaves contain the volatile principle Vellarine. Chief medicinal uses are as diuretic, blood purifier and local stimulant.
15.	Ipecac	*Cephaelis ipecacuanha* Rubiaceae	A perennial exotic (Brazil) plant, requiring cool climate, containing alkaloids in all parts, maximum being in roots (2.5 to 2.7 %), of which the non-phenolic alkaloid (emotine) is used for a number of medicinal purpose.

16.	Safed Musli	*Chlorophytum borivilianum* Liliaceae	The roots are well known tonic. It acts as an aphrodisiac besides curing diabetics, impotency, natal and post-natal problems. The major component is saponins (2-17 %).
17.	Pirandai	*Cissus quadrangularis* Vitaceae	Leaves, stem and juice of stem are used against irregular menstruation, scurvy beaten, paste given for asthma and digestive troubles
18.	Costus T: Kuroram H: Keu	*Costus speciosus* **(Koen) Sm. P.** Zingiberaceae	A rhizomatous plant, found growing wildly in tropical rain forests of Tamil Nadu and Kerala, rhizome contains 1.0 to 1.5% diosgenin, and duration is 16-18 months when cultivated.
19.	Datura **T.** Umattai H. Sadahdhatura	*Datura metal* (L.) *D. stramonium* (L.) *D. innoxia* (Mill.) Solanaceae	Annual or perennial herb or shrub found growing mostly as a .garden land weed, fruits contain alkaloid (hyoscyamine and or scopolamine), chiefly used as pre-anaesthetic in surgery and child birth and in ophthalmology.
20.	Cork-Wood tree	*Duboisia myoporoides* *D. leichhardtii* Solanaceae	Both are trees, often growing to a height of 14m. Native of Australia leaves contain 2.4 % of the total alkaloids, of which nearly 60 % is hyoscine and hyoscyamine. It is the main source of tropane alkaloids in the world today.
21.	Eclipta **T:** Karisalanganni H : Bangra	*Eclipta alba* (Hassk) Compositae	An erect, prostrate, much branched annual weedy species found throughout India, used as emetic, purgative, tonic, alterative and applied externally as antiseptic to ulcers and wound to the cattle.

22.	Buchwheat	*Fagopyram esculentum* (Hoench) *F. tataricum* (Gaertn) Polygonaceae	These are pseudo cereal, used as food grain in the Himalayas and in the hill areas of Tamil Nadu, leaves and flowers are rich in rutin 6 to 6.5 % in *F. tataricum* and 3.0 to 4.75 % in *F. esculentum.* Rutin (Vitamin P) is useful in the treatment of capillary fragility, retinitis and also for rheumatic fever of haemorrhagic conditions.
23.	Gymnema T: Sirukurinja H : Merasingi	*Gymnema sylvestre* (R. Br.) Asclepiadaceae	A climber found wildly in Western Ghats, leaves are useful for diabetes, when chewed reduces glycosuria. Roots are emetic and expectorant.
24.	Indian Saraparilla T: Nannari	*Hemidesmus indicus* (L) Shult Asclepiadaceae	It occurs in upper Gangetic plains, in all hilly areas in South India. A slender, laticiferous twining or prostrate or semi erect shrub with aromatic roots, root is the economic part, used as demulcent, diuretic, tonic, blood purifier and also against insect/snake bites.
25.	Black Henbane T: Kurunai-omam H: Khurasani-ajvayam	*Hyoscyamus niger* (L) Solanaceae	A biennial plant growing to a height of 1.6m, foliage and -seed contain tropane alkaloids of which hyoscyamine and hyoscine are important and find use as sedative.
26.	Kaempferia T: Kacholam H: Chandramula	*Kaempferia galanga* (L) Zingiberaceae	An aromatic herb, with tuberous rhizome, found throughout India, tubers are used as expectorant, diuretic, carminative,
27.	Leucas T: Tumbai H: Chola-nalkusa	*Leucas aspera* (Spreng.) Labiatae	A small herb, found more or less throughout India in the plains, used as antipyretic, insecticide, leaves are useful in chronic rheumatism.

28.	Chamomille	*Matricaria chamomilla* (L) Compositae	An herbal plant, found growing in Punjab and Upper Gangetic plains, flowers are stimulant, attenuant, carminative, oil obtained from it is used externally in rheumatism.
29.	Noni T: Nuna, H: Aach	*Morinda citrifolia* Rubiaceae	A small evergreen tree or shrub, the fruit contains more than 150 essential nutraceuticals in their most absorbable form to enhance our health, Presence of Scopoletin helps in reducing the hypertension, cures heart diseases and stroke.
30.	Phyllanthus T: Kilaanelli H: Jar-amla	*Phyllanthus niruri* (L) Euphorbiaceae	A small herb, found throughout the hotter parts of India, even in hills upto 900 m, fresh root is the excellent remedy for jaundice, infusion of young shoots is given against dysentery.
31.	T:Thippili, H: Pipul	*Piper longum* (L) Piperaceae	A creeping and rambling aromatic climber found growing in the evergreen forests of Western Ghats, dried unripe fruits are useful as alterative, tonic, and decoction of immature fruit and root- used in chronic bronchitis, cough, and cold. KAU, Thirussur released a high yielding cv. Viswam.
32.	Asoka tree	*Saraca indica* Leguminosae	Bark is the economic part, used in uterine infections and cure for scorpion sting, astringent
33.	Night-shade plant T: Manathakkali H: Makoi	*Solanum nigrum* (L) Solanaceae	A herb, found throughout India up to 2500 m even in Himalayas, berries are useful in fever, diarrhea, eye disease, juice of plant is diuretic, given in chronic enlargement of the liver, piles, dysentery etc.

34.	Stevia T:Seeni Tulsi	*Stevia rebaudiana* Asteraceae	Annual/perennial herb, the leaf extracts used to sweeten beverages like coffee, tea etc. leavs are used against hypertension, digestive disorders, cardiac- tonic properties.
35.	Strychnos T: Etti H : Kuchla	*Strychnos nux vomica(L)* Apocyanaceae	Found growing in the forests of South Indian hills; Leaves, barks, wood and root contain an alkaloid Strychnine, powdered root bark is used for treating against cholera, leaves applied as poultice to care the wounds and ulcers especially in cases where maggots have formed.
36.	T: Tuthuvalai	*Solanum trilobatum* (L) Solanaceae	A much branched climbing shrub, distributed in South and N.W. India. Berries and flowers are given in cough; decoction of the plant is useful against chronic bronchitis.
37.	T: Kadukkai H: Harir	*Terminalia chebula* Retz Combretaceae	A huge tree, found growing wildly in South Indian hills upto 1500 m and North India upto 300 to 900 m, fruit is highly astringent due to the presence of chebulinic acid, used as laxative lterative dentifrice; considered useful in carious teeth, bleeding and ulcerations of the gum
38.	Tylophora T: Asthmakodi H: Antamal	*Tylophora asthmatica* (W. & A.) Syn **T.** *indica* (Burm.6.) Asclepiadaceae	A branching climber, found in all districts of Tamil Nadu and also in states like Bengal, Orissa, etc, it contains alkaloids like tylophorine (0.1 %) tylophorinine and tylophorinidine used as a stimulant, emetic, cathartic, expectorant, stomachic and diaphoretic and in bronchitis, whooping cough.

39.	Vertex T: Vennochi H. 'Nirgandi	*Vitex negundo* **(L)** Verbanaceae	A large aromatic shurb, found grown, in all warmer parts of India, used as dimetic, expectorant, anthelmintic, febrifuge, tonic and as a demulcent in dysentery.
40.	Ashwagandha T: Amukhira Kilangu H: Asgand	*Withania somnifera* Dunal Solanaceae	A small shrub, grown in Madhya Pradesh and Rajasthan, the economic part is the dried root, contains alkaloids withanine and somniferine, used in Ayurvedic and Unani preparations against bronchitis, ulcers, rheumatism, dyspepsia etc. College of Horticulture Mandsaur (JNKVV, Jabalpur) has released two varieties viz Jawahar Asgand-20 and 134.
41.	Tea tree	*Melaleuca alternifolia L.* Myrtaceae	A native of Eastern Australia, small ever green tree, usually with single trunk, leaves contain the oil (2-2.3 %), major compound is terpinen -4 -ol which has anti bacterial, anti fungal, anti viral and anti inflammatory properties and is a broad spectrum disinfectant.

Note: T-Tamil, H-Hindi

PART-IV
AROMATIC PLANTS

22

Introduction

Aromatic plants are defined as those plants which possess **essential oil** in them. These essential oils are the odoriferous steam volatile constituents of the aromatic plants. They are mainly a complex mixture of acyclic and or cyclic monoterpenoids. These terpenoids are basically secondary metabolites and they have no apparent function in the plant's primary metabolism. These essential oils are used in perfumery, cosmetic and pharmaceutical industries whereas the essential oils obtained from spices and condiments which impart the flavour and improve the taste of the food are used in several flavour industries. Some of the special attributes of these plants are:

1. They are usually present in the aerial parts of plants such as flowers, fruits and leaves. Occasionally, they also accumulate in roots and wood (eg. sandal wood).

2. The nature and proportion of the various monoterpenoids in the essential oil is a characteristic not only of the genus but also of species as well.

3. Most of the commercial essential oil bearing plants belongs to the families Labiatae, Myrtaceae, Rutaceae, Compositae, Rosaceae, Umbelliferae, Graminae and Pinaceae.

4. The essential oil accumulation in a plant depends upon the developmental stage of the concerned organ/plant part. In most of the aromatic plants, it is associated with early growth period.

5. The composition of the essential oil also varies greatly with the developmental stages of the plants eg. in *Mentha arvensis* young leaves contain more of manthone while matured leaves contain more of menthol in the oil.

6. Essential oil content and its composition are influenced by climate or season. In geranium, for example the oil content is more (0.09 to 0.12 %) during summer months (April to June), and low (0.06 to 0.07 %) during winter months.

Some of the important features of essential oils industry in India are:

1. Current production of essential oil is about 16,000-18,000 tonnes in India as against the world production of 80,000 tonnes (i.e. 20-25 %).

2. The turnover of all perfumes in India is about Rs. 3000 million annually while the market size of Indian cosmetics is around Rs. 4,5000 million. Market growth in India is expected to be around 15 percent compared to 5-7 % in America or Europe.

3. India ranks 28th in import and 14th in export of global trade in essential oil.

4. India is the largest producer of Mints and basils.

5. China, Brazil and Indonesia are stiff competitors for India in this industry.

Aroma therapy is a form of alternative medicine that uses volatile plant materials and other aromatic compounds for the purpose of altering a person's mind, mood, cognitive function or health. Nowadays, it is gaining importance among the tourist and rich segment of society and its overall benefit to human health. World consumption in aroma therapy is still small (200-300 tonnes) but is expected to grow if combined with the pharmaceuticals.

Distillation of Essential Oil

Essential oils are complex mixtures of odoriferous and steam volatile compounds which can be separated by distilling with water. This process is known as **hydro distillation.** There are three types of hydro-distillation.

1. **Water distillation:** This is the simple method in which the plant material to be distilled comes in direct contact with boiling water in a distillation still. This method is advantageous for certain materials especially when they are in finely powdered form. But it is not good for material containing saponifiable or high boiling point constituents.

2. **Water and steam distillation:** This is an improved method in which the plant material is supported on a perforated grid or false bottom above the bottom of the distillation still. The lower part of the still is filled with water to a level below the false bottom. When heated, the wet steam of rises through the material at low pressure. In this method, hydrolysis is fairly at a low rate, while the distillation is rapid, the oil yield is also better and the physico-chemical properties of the oil are also good.

3. **Direct steam distillation:** This method is similar to resemble the second method except that no water is kept in the bottom of the still. Live steam saturated or super heated steam with pressure higher than atmospheric is passed through the bottom of the still. The rate of distillation is very high and any yield and quality of oil are also good in this method. In this process, steam does not penetrate the cell membranes and the essential oil is vaporized only after diffusing out as an aqueous solution through the cell membranes. A steam distillation unit (Figure 21) consists of the parts like (1) still (2) condenser (3) separator and (4) steam generator still serves as a container for the plant material to be distilled and is usually cylindrical/ vertical tank equipped with a removable cover which can be clamped upon the cylindrical section. There is a grid or false bottom nearby the bottom on which the plant material rests and live steam is introduced below the grid. The condenser serves to convert the steam accompanying oil vapours into liquid. An efficient type is of multi-tubular condenser. Condensation is achieved by the circulation of cool water. The function of oil separator is to achieve a quick and complete separation of the oil from condensed water. Volatile oil and water are mutually insoluble. These two liquids form two separate layers due to the difference in their specific gravities. Since the total volume of water condensed is always greater than the quantity of oil, water is removed continuously.

Importance of Essential Oil Industry in India

1. Use of aromatic plants and their products is as old as our history that distillation of rose flowers is mentioned in *Charaka* and *Sushrita Samhitas* believed to have been written in Indian 1000 years B.C.

2. The aromatic plants and aroma chemicals contained in them play a vital role in our day to day living. More and more common and middle class people are using perfume and perfumed products, which

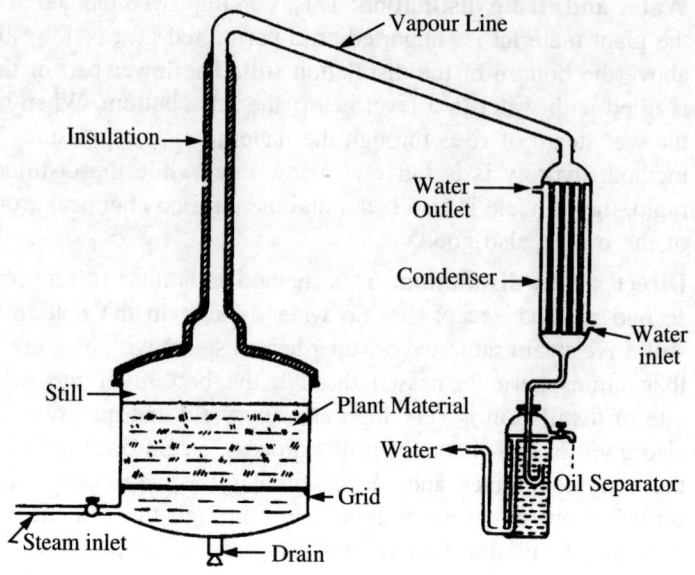

Figure 21 A steam distillation unit.

were previously used by affluent and rich people only as these are falling with the reach of more and more.

3. India has varied climate conditions and suitable soil exists in one or other part of the country. Hence it must be possible to grow almost any type of essential oil bearing plant (Figure 22). Historically, India has enjoyed a pre eminent position as the supplier of natural perfumes to the world over. This is still true in the case of sandal wood oil, lemongrass oil, palmarosa oil, vetiver oil and cedarwood oil.

4. Though more than a thousand of Indian flora have been reported to contain odoriferous materials, only half a dozen have been systematically studied and cultivated.

5. Our country is earning a foreign exchange of Rs. around 130 crores per annum by way of export of aromatic essential oil. However, its contribution in world export is only 1.1 % and in import 0.7 %.

6. India ranks 28th position in import and 14th position in global trade of essential oil.

7. India's share of essential oil in the world market can be improved greatly if some of the bottlenecks that prevail now are removed. They are

(a) Adoption of age old technology is still being followed in essential oil production.

(b) Wide fluctuation in quality and price.

(c) Availability of low priced synthetic substitutes.

The cultivation details of selected commercially important aromatic plants are dealt in chapter 23 while a brief description of other important aromatic plants is mentioned in chapter 24. An account of floral concrete and other aromatic products obtained from commercial flower crops are also dealt in chapter 24.

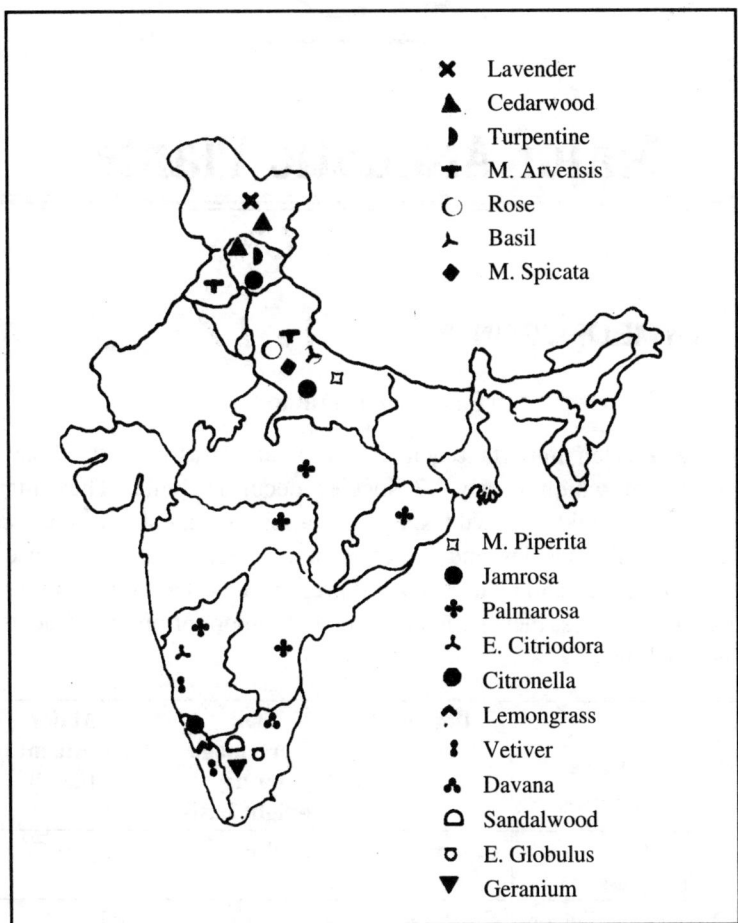

Figure 22 Important aromatic plants grown in India.

CHAPTER

23

Major Aromatic Plants

23.1. CYMBOPOGON SP.

Family: **Graminae**

Cymbopogon is an important genus of aromatic grasses with about 120 species and of which nearly 27 species occur in India. They include cultivated, semi wild and wild species. Various essential oil constituents are present in the species and varieties of this genus which are used in perfumery, cosmetic and pharmaceutical preparations. The commonly grown Cymbopogon species, their popular name and important oil constituents are presented below:

	Common Name	Botanical Name	Oil percentage on dry weight basis	Major constituent in the oil
1.	East Indian 1 Lemon grass	*C. flexuosus*	1.0 to 1.2	Citral (80 %)
2.	West Indian Lemon grass	*C. citratus*	1.0	Citral
3.	Jammu or North Indian Lemon grass	*C. pendulus*	0.75	Citral (75-80 %)

4.	Ceylon citronella grass	*C. nardus*	0.70	Citronellal (20-25 %)
5.	Java citronella grass	*C. winterianus*	1.2 to 1.5	Citronellal (30-38 %)
6.	Palmarosa	*C. martinii* var. *motia*	0.43 to 0.52	Geraniol (95 %)
7.	Ginger grass	*C. martinii* var. **sofia**	0.75	Geraniol (35 to 50 %)
8.	—	*C. khasianus*	1.6 to 2.2	Citral (40 to 60 %)
9.	—	*C. nardus* var. *confertiflorus*	0.8 to 1.0	Geraniol (70 %)
10.		*C. jawarancusa*	0.8 to 1.0	Piperitone (15 to 20) and Linalool (20 to 25 %)

In India, lemongrass, citronella and palmarosa grasses are cultivated to a greater extent and their cultivation details are discussed in this chapter.

23.1. LEMON GRASS

Three types of lemon grasses viz. East Indian lemongrass *(C. flexuosus)*, West Indian lemon grass *(C. citratus)*, and Jammu lemon grass *(C. pendulus)* are in cultivation in our country as the important sources of cirtral.

C. flexuosus **is** grown commercially in Kerala and nearby adjacent states. Its oil is popularly known as "Cochin oil" as it is shipped mainly through Cochin port. India is annually producing nearly 1000 MT per year while the world demand is much more. Annually, we are exporting lemon grass oil to a tune of about Rs. 5 crores. Our country is facing a critical competition from Gautemala in the International market.

Uses

1. The chief constituent of the oil is the citral; it is the starting material for the preparation of important ionone *viz.* α-Ionone - used in flavours, cosmetics and perfume and (β-Ionone - used in the manufacture of synthetic vitamin A.

2. Oil has bactericidal, insect repellent and medicinal uses.

3. The spent grass is a source of good cattle feed and can be converted into good silage.

4. Spent grasses are also useful for the manufacture of cardboards and papers or as fuel.

Climate and Soil

The plants are hardy and grow under a variety of conditions. The most ideal conditions are a warm and humid climate with plenty of sunshine and rainfall of 250-280 cm per annum, uniformly distributed. Regarding the soil, it can be grown from poor soils, in the hill slopes. Soil p^H ranging from 4.5 to 7.5 is ideal. As it has good soil binding nature, they can be grown as vegetative cover over naked, eroded slopes.

Varieties

Name of the variety	Institute	Characteristics
Pragathi	CIMAP, Lucknow	Dark purple leaf sheath and citral content 85-90 %, average oil content 0.63 %
Nima	CIMAP, Lucknow	Tall, citral type
Cauvery	CIMAP, Lucknow	Tall, citral type
Krishna	CIMAP, Lucknow	Medium tall, high tiller and herb with high oil yielding citral
NLG- 84	AICRP on M & AP, NDUAT, Faizabad , UP	It is a selection from Germplasm. High herbage yield, 100-110 cm tall, broad leaves with dark purple sheath. oil content – 0.4 %, citral content – 84 %.
Sugandhi (OD-19)	AMPRS, Odakkali, KAU, Kerala	Compact plants with red stemmed variety, thin green leaves, oil yield – 100-125 kg/ha, geraniol 84-86 %.
SD-68	CIMAP, Lucknow	Developed by ionizing radiation, yields upto 375 kg of oil/ ha/yr with a citral content 90-92 %
Chirharit	G.B.Pant University, Pantnagar	High yielding variety, developed by systemic breeding for genetic improvement. It is frost resistant and the essential oil contains 81 % of citral
RRL 16	IIIM, Jammu.	Evolved from *C.pendulus* ,average yield of herp is 15 to 20 t/ha/annum giving 100 to 110 kg oil, oil content varies from 0.6 to 0.8 % and citral content is 80 %.

North Indian Lemon Grass (*Cymbopogan pendulus*)		
CKP25	IIIM, Jammu.	An interspecific hybrid between *C. khasianus* × *C. pendulus*. Citral content of oil ranges from 80-85 %, capable of yielding 50 % and 140 % more oil yield than RRL-16 and OD-19 respectively.
Kalam (CPK-F$_2$-38)	IIIM, Jammu.	It is an improved F$_2$ derivative of F$_1$ hybrid of *C. pendulus* × *C. khasianus*. Average essential oil yield is 170-190 kg/ha/yr. Citral content is 75 -85 %

Nursery

The soil should be well pulverised for forming the seedbed and it should be a raised bed one. Leaf mould and farm yard manure are also added to the soil while forming the bed. 15-20 kg of seeds is required for raising seedlings for one hectare. Seeds are sown in lines drawn at 10 cm interval in the beds and covered with cut grass materials. When the seedlings are about 2 months old or about 12 to 15 cm high, they are ready for transplanting.

Field Preparation

The land is cleared of the underground vegetations and pits of 5 cm cube are made at a spacing of 15 × 10 cm. Splits from old clumps can also be used for propagation.

Manures and fertilizers: The Aromatic Plants Research Station Odakkali (Kerala) recommends 100 kg of N/ha. Under North East conditions, application of 60 kg of N, 50 kg p and 35 kg K is recommended per hectare.

Harvesting

Lemon grass comes to harvest 90 days after planting and subsequently it is harvested at 50-55 days interval. The grass is cut 10 cm above the ground level and 5-6 cuttings can be taken in a year subject to the climatic conditions. Depending upon the soil and climatic condition, the crop can be retained in the field for 5 to 6 years.

On an average 25 kg of oil can be obtained from first year per hectare plantation and about 80 to 100 kg of oil per year from 2nd to 6th year if well maintained.

In North India, Jammu lemon grass (**C. pendulus)** is cultivated under irrigated condition as a source of citral. The cultural practices are almost similar to East Indian lemon grass. It is propagated exclusively by slips which are planted on flat beds. A spacing of 50 × 50 cm is adopted. A dose of 260 kg N, 80 kg P_2O_5 and 120 kg K_2O per hectare is recommended in 3-4 split doses. The crop responds to irrigation especially during hot summer months.

Depending upon the planting period, one or two cuttings are taken in the first year and from second year onwards, 3-4 cuttings are available. Harvesting consists of fresh leaves and also the dry or semi-dried leaves at intervals of 60 days. The crop should not be allowed to flower profusely as it reduces the over all oil yield.

Two improved varieties available in *C. pendulus* are as follows:

Name of the variety	Origin	Herb yield (t/ha)	Oil yield (kg/ha)	Oil content (%)/dry basis	Proportion of citral (%)
RRL-16	IIIM,Jammu	55–65	120–150	1.5–1.7	75–85
Praman (Tetraploid)	CIMAP, Lucknow	80–85	230	2.5–3.0	82–85

23.1.2. CITRONELLA GRASS

Java citronella *(C. winterianus)* and Ceylon citronella *(C. nardus)* are the two primary sources of citronella oil. Among these, the Java type is considered to be the best source of citronella oil and its cultivation has now become popular. It is chiefly grown in Formosa, U.S.A., U.K. France, West Germany, Japan, China and Java. In India, it is grown mostly in Assam and to a limited extent in states like Uttar Pradesh, Andhra Pradesh, Karnataka, Gujarat and Tamil Nadu. The area under this crop is estimated to be around 4500 ha and the demand for this oil is on the increase in view of its use in perfumery and cosmetic industries.

Climate and Soil

The plant prefers a warm climate with plenty of sunshine and an annual rain fall of 200-250 cm spread over a greater part of the year. Prolonged drought is harmful for growth and development. It does not grow happily under shade as shade results in poor growth. It grows well in the plains and also in hills up to 900 m but the best elevation is 180 to 250 m MSL. Places having 75 to 90 percent relative humidity favour the growth always and the moisture in the atmosphere is more important than the moisture in the soil.

Citronella grass is a moisture loving plant but prolonged water stag nation in the soil is undesirable for the growth of the plant. Rich sand loamy soil is most suitable while heavy and loamy soils of poor texture are not suitable. It grows within a p^H range of 5.0 to 6.5 but in Gujarat it grows well in neutral soils.

Varieties

The Regional Research Laboratory at Jorhat has developed an improved strain (RRL - JOR-3-1970) which is giving 20-25 percent high leaf yield than that of the local types. Citronellal content is 46.2 %. The NBPGR, Delhi has also identified two improved strains viz IW 31243 and I.W. 31245. Bidhan Chandra Krishi Viswa Vidyalaya, Kalyani (W.B) has also developed selection viz KS-CW- SI. CIMAP, Lucknow has developed the following varieties in Java Citronella (*Cymbopogon winterianus*).

Name of the variety	Characteristics
Manjusha	Medium tall herb, high tillers and high oil yielding
Mandakini	Medium tall, high tillers and herb with high oil yielding
Manjari	Elite mutant clone of Manjusha , medium tall herb, high tillers, rapid growing ability, dark purple stem and high herb yield, high citronellol and low elemol content
Bio 13	Somoclonal variety, medium tall herb, high tillers and high oil yielding
CIM Jeeva	High field establishment capacity, suitable for north plain establishment capacity > 90 %
Medini	Very tall, geraniol type (60 %)
Ceylon Citronella (*Cymbopogon nardus*)- IIIM, Jammu	
RRL-CN-5	A perennial densely tufted grass attains a height of 100-150 cm. Essential oil yield : 280 kg/ha Geraniol – 60 to 70 %, Genanyl acetate – 10 to 20 %
Jamrosa	Interspecific hybrid *C. nardus* var. *confertiflorus* X *C. jwarancusa*. the oil is good substitute for palmarosa oil with higher yield/ha. Withstand water stress condition.

Preparation of Land

Land should be prepared well and beds and channels are formed in the plain. In hills, the lands are cleared of under ground vegetations during the dry months and kept ready for planting. Rooted slips are used as the planting materials and planting commences during April/May after the onset of monsoon. A spacing of 90 cm between rows and 60 cm within the row is followed and two slips are planted per hole taken to a depth of 15 cm.

Manures and fertilizers: Citronella grass is a soil exhaustive plant and it requires liberal application of fertilizers. A basal dose of 62 kg of P and 50 kg K_2O per hectare per year are applied at the time of planting or first cutting in the beginning of every year and 125 kg of N is applied in equal split doses one month after every cutting. It also responds to foliar spraying of nitrogen as one percent urea spray at fortnightly intervals during the growing period of the year. Sometime, the plants exhibit deficiency symptoms of micronutrients especially Zn, Fe and Cu which can be corrected by spraying 0.2 % Zn, 0.6 % Fe and Cu 1 % along with the Urea spray or individually.

Harvesting

Generally, the plants grow at slower rate till five months after planting and thereafter, they start growing at a faster rate. First cutting may be ready 4 to 5 months after planting and thereafter cuttings are taken at 2 months interval before winter period. If irrigation is given during dry periods, one or two more cuttings may be obtained from November to February. Ideal harvesting technique involves cutting the leaves just above the growing point. Once in a year the clumps should be cut close to the ground level to remove the dry leaves and maintain the bush height to a manageable size. The harvested grass can be semi dried for 48 hours before distillation of oil.

Semi-dried grasses are chopped before distillation. Steam distillation gives a good oil recovery of 1.0 to 1.2 percent. Distillation time runs to 90 to 105 minutes per batch.

The yield of oil varies with the season, fertility of soil, method of distillation etc. Though the plantation can be kept for 5 years, the yield of oil will be normally high during the second and third year and thereafter it declines. From a well maintained plantation, the following oil yield can be obtained:

First year	– 140-150 kg/ha
Second to third year	– 200-300 kg/ha
Fourth year	– 200 kg/ha
Fifth year	– 100 kg/ha

The spent grass will be useful as a mulch material as well as an organic manure for the same field. It can be also used for making paper, paper boards etc.

Apart from citronellal (32–45 percent), this oil also contains geraniol (12–18 %), citronellol (11–15 %) and geranyl acetate (3–8 %) which can be converted into many other products as synthetic menthol, esters of geraniol and citronellal etc.

23.1.3. PALMAROSA GRASS

Oil of palmarosa, also known as Rusa or Rusha or Rosha is the essential oil obtained from the aromatic grass *Cymbopogon martinii var. motia.* The oil obtained from the other form viz. C. *martinii* var. *sofia* is known as the ginger grass oil. Motia and sofia grasses are almost identical and difficult to distinguish when they are in earlier growth stages. Motia grass has fine yellow stem with dark green leaves and they attain a height of 1.80 - 2.40 m. On the other hand, sofia grass has purple stem, shorter (90-120 cm) than the motia grass. These oils are used as base for the fine perfumery and are valued because of their geraniol content. The oil is useful in imparting rose - like aroma to a wide variety of soaps, tobacco products etc. The oil of palmarosa is commercially preferred to ginger oil.

Climate and Soil

Palmarosa is a hardy plant and can grow in varying altitude right from sea level. It stands well in places receiving rainfall from 75 cm to 150 cm. But it does not withstand stagnant water. It requires exposed sunlight and does not perform well under shady situations.

Palmarosa prefers a well drained soils of neutral to alkaline reaction and can be grown in poor sandy to heavy fertile soils of arid tracts, saline soil conditions and also in marginal and sub marginal lands.

Varieties

Variety	Institute	Characteristics
Rosagrass Haryana-49 (RH-49)	AICRP on M&AP, CCS HAU, Hissar, Haryana	It is a clonal selection variety. Perennial crop – 3 to 4 years. Oil yield/annum – 75 to 100 t/ha (2-3 cuts) Geraniol content 90 %.
Trishna	CIMAP, Lucknow	Plant is tall herb having high tillers and long inflorescence. Oil yield is high.
Tripta		Yield potential 320 kg oil/ha/year.
PRC -1		Tall, high tillers and herb with high oil yielding.

Composite-12		Tall, high tillers and herb with high oil yielding.
Vaishnavi and Harsha		—
Jawahar Rosa 68 (JR-68)	JNKVV, College of Agriculture Indore Centre, MP	Dry herb yield -150 to 160 q/ha, oil content : 1.01 % Total gernaiol content – 91.7 %
CI-80-68		175-200 cm tall, medium tillering capacity; essential oil content 1-15 %
Sel. IW-31243 and IW- 31245	NBPGR, New Delhi	—
RRL(B)-77 and 71	RRL, Bhubaneswar	—

Land Preparation

Nursery bed should be prepared out of well pulverised soil and at a raised level. Leaf mould or farm yard manure should be mixed well with the nursery bed. Seed rate is 2.5 Kg per hectare. Best time for sowing is from April to September. Seedlings become ready for transplanting when they are about 15 cm high.

The main field for raising palmarosa should be prepared by ploughing 3-4 times, followed by forming ridges and furrows at 90 cm apart. The seedlings are transplanted at 60 cm spacing in the ridges. In Kerala, a spacing of 45 × 30 cm is followed but under Delhi condition, a closer spacing of 45 × 15 cm is found to be good in producing higher herbage and oil yield. In North, it is recommended as a mixed crop along with Basil to get higher net profit from unit area.

Manures and fertilizers: It responds to application of compost. Under North Eastern condition of India, a fertilizer dose of 60:40:40 NPK kg/ha is followed while under Kerala condition, NPK dose of 25, 50 and 25 kg/ha is recommended. Micronutrient like zinc ($ZnSO_4$ 25 kg /ha) is beneficial to increase the oil yield of palmarosa.

Harvesting

If transplanting is done in May/June, the grass comes to first harvest after six month from transplanting. Harvesting consists of cutting the upper third of the stem along with the leaves. The right time for harvesting is when the plants just begin to bloom as the leaves contain higher oil content during the blooming period. Recent studies at NBPGR, Delhi shows that oil from the whole plant is of good quality and economical to produce as per the table below:

	Plant part	Essential oil (fresh weight)	Geraniol (%)
1.	Whole plant	0.53	72.4 to 86.5
2.	Inflorescence	0.83	72.5 to 84.5
3.	Leaves	0.58	92.9 to 94.6
4.	Stalk	0.01	–

Besides, superior quality of oil with roseous green odour is obtained at an early seed setting stage rather than at full bloom stage when the oil yield is slightly higher.

The grass yield more oil recovery if dried for nearly one week. The steam distillation seems to be better than the other type of extraction. Palmarosa yields 15 to 20 tonnes of herbage per ha in a year with an oil yield of 50-60 kg per year. The plantation can be maintained for about 8 to 10 years, but the oil yield starts declining from the fifth year.

Other Sources of Geraniol Rich Cymbopogon Species

The oil of palmarosa contains about 95 % geraniol. The oil of ginger grass contains 36 to 65 percent of geraniol, hence used as an adulterant for oil of plamarosa. Recently IIIM, Jammu has developed a hybrid viz Jam Rosa (RRL-82), a good geraniol rich *Cymbopogon*. It is a hybrid between C. *nardus* van *confertiflorus* and C. *jawarancusa*. The plants are more bushy having more leaves smaller internodes, and more number of tillers per clump. Its oil content is 1.0 to 1.2 % with a total geraniol content of 75-85 %. This hybrid gives an essential oil similar to the palmarosa oil of commerce. The other geraniol rich *Cymbopogon* varieties developed are:

	Varieties	Species	Oil (%)	Total geraniol (%)
1.	RRL-38	C. *flexuosus*	1.2 to 1.5	50 to 60
2.	RRL-59		1.0 to 1.2	45 to 50
3.	Cl-269		0.50	84.9
4.	RRL-14	C. *khasianus*	0.17 to 0.19	77-78

23.2. VETIVER *(Vetiveria zizanioides* L.)

Family: **Graminae**

The vetiver grass (***Vetiveria zizanioides***) is the source of the valuable aromatic vetiver oil which enjoys world-wide reputation being one of the finest oriental perfumes. It is called 'Khus' in Hindi and Bengali, Vetiver or Vela-machamver in Tamil. As an important essential oil-yielding plant of India, vetiver has been grown and extensively used in the country for several centuries. India has been exporting vetiver oil worth more than 1.5 lakhs but still there is much scope to increase the export. Its medicinal and commercial importance was known even in the earliest days of Indian civilization. The vetiver grass is a native of India and is found through out the plains and lower hills (upto 1220 metres in altitude) of India, Burma, Ceylon. It is systematically cultivated in Kerala, Tamil Nadu and Andhra Pradsh.

Uses: Roots of vetiver are used for a variety of purposes. The bulk of the oil produced in India is consumed for soap making. Apart from extraction of oil, the roots are used to make aromatic mats, brooms, screens, pillows and mattresses along with bamboo splits. In summer seasons a few bits of roots are put in drinking water for imparting the aroma.

The roots are also used in pharmaceutical preparations both in the Allopathic and Ayurvedic systems of medicines. It is used as a stimulant, refrigerant and stomachic.

The vetiver belongs to the genus ***Vetiveria*** which includes seven species. It is a perennial grass, densely tufted upright, often growing luxuriantly on rich, marshy soil, attaining a height of 1-1.8 metre with root portion branching into spongy, aromatic and fine rootlets. It grows in large clumps, the leaves of which are long, erect, narrow, stiff with high margins and non-aromatic and upto one metre in length.

The grass puts forth a long terminal panicle carrying numerous slender racemes of spikelets. The awnless, spikelets are in pairs, one sessile and perfect, the other pedicelled and staminate, the sessile spikelet bearing minute spines. Many of the cultivated types rarely flower and the flowers that are seen on other types do not get seeds.

It is the root which gives the essential oil, and is strongly scented. The length of the roots varies from 10 to 35.5 cm and some times even more.

Varieties

There are basically two main types of the vetiver grass, viz.

1. Seeding type and 2. Non-seeding type. The one that grows wild in North India is mainly the seeding type while that of the South is the non-seeding type. Besides, differences exist between the North and South Indian strains with regard to yield and aroma of the oil. The oil of some of the North Indian types viz. Bhartapur, Akil and Musanagar strains generally have an aroma superior to that of oil derived from South Indian grasses.

Crop improvement studies through selections and hybridizations have led to identification of superior strains. The Indian Agricultural Research Institute, New Delhi has identified three selections which give nearly double the root yield than that of local types. Similarly, some superior hybrids viz. Hybrid 8 and Hybrid 7 developed at the IARI were found to be superior in terms of root yield and oil yield. In Tamil Nadu, the Hybrid 7 was found to be performing equal to the local types. In Kerala, Hybrid ODVI 3 has been identified as a superior variety than the local Nilambur type (3.8 t/ha roots (fresh weight), 15-22 kg oil). CIMAP, Lucknow has developed the following high yielding varieties:

Variety	Characteristics
Dharini	Khus note, very tall, light red/pink coulor of inflorescence, high tiller, Dense long root network, a suitable soil- binder –cum-high oil yielding.
Gulabi	Has rose note yields 28q/ha of dry roots, 25-30 kg essential oil.
Keshari	Very early flowering, greenish white inflorescence, medium long roots with high oil yielding.
CIM Vridhi	Khus notes, medium tall, late flowering, red colour of inflorescence
KS-1 and KS-2	Quality of oil from both the clones are the best.
Sugandha	Tetraploid strain, 1.4% oil content in fresh roots, 21.2 q roots/ha

Climate and Soil

Vetiver prefers a mild climate but can be grown under both wet and dry or arid tropical conditions. Under temperate or warm winter hill areas, the growth of vetiver remains stunted.

The most suitable soil consists of loose sandy soils, preferably on the sloppy hills. In such soils only, the roots can be easily pulled out without much loss of thin roots. Compact and heavy soils may be avoided.

Planting

There are three different systems of planting adopted by different growers.

System I-conical ridges, 30-38 cm high and 48 cm apart are made at the summit and the slips planted 23 cm apart on the summit.

System II-the land is laid out into beds of 30 cm high, 68 cm wide and 45 cms apart edge to edge and the slips are planted on these in two rows 22.5 cm apart, leaving 22.5 cm on either side.

System III-the beds are made 45 cm high, 60 cm wide and 30 cm apart edge to edge and two rows, 30 cms apart, are planted on these leaving 15 cm on either side. The spacing within the row is also 30 cm in this system. Trials conducted in the cinchona plantations at Anamalais to determine the relative merits of first and second mentioned systems of planting have indicated that the latter method gives higher yields of roots. The cost of harvesting is also lesser in this method.

Vetiver can be propagated through tillers and slips. Tillers take long time for growing and therefore, slips are the better planting materials for propagation. The top of the slips are cut down before planting to prune transpiration loss, thus giving a better chance for survival of the slips.

The slips are planted in pits, five to eight cm deep made with a pointed stick. Two or three slips are planted in each hole. This is done to meet any casualties and also to get a thick stand of plants. After planting, the soil around the slips is pressed firmly and leveled. One hectare requires 1, 50,000 to 2, 25,000 slips with 2-3 slips per pit in the commonly adopted system of planting (IInd method).

The best planting time to get higher oil yield under South Indian condition is June-July. Vetiver should not be planted on shaded places as shade will exert an unfavorable influence on the development of root system.

Manures and fertilizers: Generally, application of groundnut cake or cattle manure has a beneficial effect on the yield of roots. Fertilizer trials conducted at Research Station, Odakkali in Kerala revealed that application of 22.5 kg P_2O_5 and 22.5 kg K_2O is optimum to get higher yield. Under North Indian conditions, vetiver clones need to be fertilized with N (20 kg/ha), to be given after two months of planting to produce higher root yield.

Harvesting

Harvesting is done during the dry months of the year. In general it is the practice to harvest the roots both for manufacture of articles and for distillation when the plants are about 10-12 months old. In Kerala, it has been established that 18 month old plantation yield economic root yield.

Harvesting or uprooting is done with digging forks having prongs of 45 cm length. To start with the stem portion is cut at a height of 15-20 cm and the clumps are then uprooted. About 50-60 per cent of the roots come away with the clumps leaving the rest in the soil. The clumps are beaten on a piece of log to remove stones and earth adhering to the roots and the roots are separated with a sharp knife. As far as possible, the ˉoots left in the soil are also collected.

The roots that possess the following characteristics have good oil content. It should

1. expose a hard surface when the skin is peeled off
2. be thick, hard, long and wiry and
3. give a very bitter taste when chewed.

The roots should not be extracted from the ground earlier than 24 month after planting if high quality oil is desired. Young roots are tender, thin, almost hair like, on pulling they break easily and stay in the ground. Besides, on distillation, they yield an essential oil with a low specific gravity and low optical rotation. The odour of these light oil is 'green', earthy'. On the other hand, older, well developed, thicker root yields an oil of better quality, with better odour and are more lasting. Oils derived from older roots are usually having a darker colour than the oil distilled from younger root.

Extraction of vetiver oil is produced by water distillation and steam distillation of the roots. In both these methods, the roots are first cleaned, steeped in water for about 12 hours, chopped into small pieces (5 to 10 cm long) and then distilled. Although steam distillation is more economical and gives of oil, yet it is not preferred by majority of the distillers. It is because of the fact, that oil produced by steam distillation does not give the same colour as produced by water distillation. Distillation time varies from 12 to 36 hours.

Yield

On an average, one hectare of vetiver plantation yields 5 to 7 tonnes of roots which on distillation yield 15 to 16 kg of oil. Roots yield 1.00 to 1.50 percent of oil on dry weight basis. The colour of the oil is light yellow and the oil contains 65 to 75% veteverol.

23.3. GERANIUM (*Pelargonium graveolens* (L.) Hervitt.)

Family: **Geraniaceae**

Geranium, a native of Cape Province (South Africa), was first introduced from Reunion island at Yercaud by a French planter, Monsieur Sens in early 20th century. It was then introduced in Nilgiris in 1954 and has since naturalised itself so well under the South Indian subtropical climatic conditions that to-day it is an excellent economic crop in many parts of India. Geranium is cultivated on a commercial scale in Reunion, Algeria, Morocco, Belgium, Congo, Spain, France and India. In India it is grown in the South Indian hills and at lower altitudes of Kashmir and Jammu. India is importing more than 18 tonnes of geranium oil from other countries at a high cost to meet the needs of Indian perfumery industries, against an internal productioin of only about 5 tonnes of oil.

The oil of geranium is widely used in many perfumery and cosmetic industries due to its agreeable and very profound strong rose- like odour. If pure, the oil is almost a perfume by itself and blends well into all kinds of scents, floral as well as oriental. It is widely used in scenting of soaps and for the isolation of rhodinal which forms part of most high grade perfumes. Tannin can also be obtained as a bye-product from geranium stems and leaves after extraction of oil.

The commercial oil of geranium is obtained from *Pelargonium graveolens*. The leaves are palmately lobed and densely pubescent. The flower is a panicle, borne on long peduncles and the flowers are light or deep pink and small.

Varieties

Two types of geranium i.e. Algerian (or Tunisian) and Bourbon (or Reunion) are being generally cultivated. The former type is slender with dark pink flowers and unsuitable for wet conditions, while the latter type is more sturdy with light pink flowers and more suitable for wet conditions. Horticultural Research Station, Kodaikanal (TNAU) has developed an improved variety viz. **Kodaikanal -1** geranium. It gives high herbage (45.2 tonns/ha) and oil yield (54.0 kg). The essential oil contains more percent of geraniol (60.5 %). **CIM-Pawan** is another improved variety developed by CIMAP which has higher oil content (0.18 %), herbage yield etc. Other high yielding varieties developed by CIMAP are **Hemanti, Bipuli and Kunti.**

Climate and Soil

Geranium can be grown in temperate, sub-tropical and tropical climate at various altitudes. However, it thrives best in a sub- tropical climate with a temperature ranging from 5.0° to 23.0°C. However, temperature below 3.0°C kills the plants. Geranium flourishes well in places receiving rains distributed throughout the year but heavy rains resulting in water-logging cause root-rot and stunt the growth. The plant itself is quite resistant to drought but not the frost. However, prolonged spells of dry weather lower the yield of oil substantially.

It is a shallow rooted crop and hence requires a well drained soil. The crop is found to perform well in the peaty and lateritic soils with a pH of 5.5 to 7.0, though a calcium rich porous soil is much desired. Reddish gravely soil and black cotton soils are also considered suitable.

Propagation

Geranium is easily propagated by cuttings. Since this crop does not set seed in our country, vegetative propagation is resorted to, as the only way of multiplication. Terminal cuttings of 10 to 15 cm long are the best suited material for propagation as they give 80 % rooting. But it has been now established that by treating the middle cuttings and basal cuttings can be also induced to root up to 80 % and 65 % respectively by treating with regulators like IBA or IAA at 200 ppm for 6 minutes.

The cuttings are planted in raised beds of 3 m long and 1 m wide. The best pH in the soil is 5.0 to 5.5 for maximum rooting. The soil is mixed well with powdered and well rotten farmyard manure. The cuttings are planted closely with spacing of 8 to 10 cm. The cut ends of stem cuttings are dipped in 0.3 % solution of wet cerasan or benlate. A temporary shade is provided until root initiation. The beds are watered judiciously. The cuttings will be ready for planting in 60 days. Cuttings can also be rooted in polythene bags or with a ball of moss so that they can be planted in the field without disturbing the root system. Such a practice will ensure high percentage of survival in the field.

Cultivation

Geranium is grown either as a pure crop or as an intercrop in the orchards. The land selected should preferably have a gentle slope and should be sheltered from heavy winds. The field should be given preliminary digging. Contour trenches of 15 cm depth are made and the basal dose of farmyard manure and fertilizers are applied and incorporated in the soil. Only well rooted cuttings are planted in the field with a spacing of 60 cm × 60 cm on the contour trenches during June-July after the commencement of the South West monsoon when the soil is fully moist. About 30,000 cuttings are required for planting one hectare. Recently, a closer spacing of 45 × 45 cm accommodating 57,500 number of plants/ha is also found to give high herbage yield. The plants will establish well in the field in two months of planting. Mulching the field with pine needles conserve the soil moisture and suppress the weed growth initially. The field has to be kept free from weeds and wild growth by periodical hoeing. Deep digging is to be avoided as they induce soil erosion and leaching of nutrients. Similarly hoeing during dry months has to be discouraged to conserve moisture. In general, weeding must be effected with minimum soil disturbances without affecting the root system.

Manures and manuring: Geranium responds well favourably to major and micro nutrients. A basal dressing of 25 tonnes of farm yard manure, 75 kg of urea, 400 kg of superphosphate and 100 kg of muriate of potash is recommended during the first year of planting.

The farmyard manure is applied before the last digging and the fertilisers are applied in the contour trenches and incorporated into the soil before planting. From the second year onwards, urea is applied @ 30 kg/ha at the end of each harvest. The 400 kg of superphosphate and 100 kg of muriate of potash are applied in a single dose during June-July. Foliar spray of urea (2%) @ 15 kg per ha given in two split doses 45 and 60 days, over a basal application of 15 kg of urea per ha is found to be more effective on the foliage. Zinc sulphate at 20 kg/ha and boron at 10 kg/ha basally once in year during June-July improves the yield of herbage. Geranium also responds well to copper and molybdenum.

Harvesting

The first cutting of the leaves can be done 8 months after planting and thereafter the crop can be harvested once in three or four months depending upon the fertility of the soil rainfall, and occurrence of frost. The plants can be kept productive for five years with proper care. All plants which show signs of wilting and decay must be removed and replanted during the monsoon season. Once in five years the bushes are cut close to the ground (10 to 15 cm) and rejuvenated.

The maturity of plants for harvest can be judged by the following symptoms:

1. The basal leaves will begin to turn yellow.
2. When the leaves are pressed between the fingers, delicate rosy odour will be emitted.
3. The lemon like odour of the leaves will change to a pronounced rose note.
4. Flowers may appear here and there.

The oil content of geranium ranges from 0.08 to 0.15 per cent and is much influenced by the stage of the crop and also the period of harvest.

The oil content is higher during summer months from April to June. High humidity and rains at the time of harvest lowers the yield of oil. The terminal portion with 6 to 12 leaves contain more oil than the middle and basal portions. The leaves below 12th node contain very little oil. Thus, the yield of oil obtained from the plant depends on the proportion of leaf surface present. Heavy, woody stems contain practically no oil and their presence therefore merely adds to the cost of harvest and distilling.

Extraction

The freshly harvested terminals are used for distillation of oil. The plant materials are stacked near the stills for about 12 to 24 hours. This results in a slight fermentation and splitting of the glycosides increases the yield of oil. The distillation process may take 4 to 6 hours depending upon the distillation unit.

The geranium oil possesses a strong rose-like odour. The chief constituents of the oil are geraniol (68-75 %) and citronellol (23-40 %). The quantity of geraniol and citronellol and their ratio vary in different geranium oils.

Yield

The yield depends mainly on the total population of the field. A minimum of 25,000 plants should be maintained in a hectare. 15,000 to 18,000 kg leaves can be harvested from a hectare in a year and they may yield 15 kg of oil on distillation.

Plant Protection

Geranium is found to be affected by tip-rot caused by *Gleosporium* spp. and root rot caused by *Fusarium* spp. during South West monsoon season. The wilted plants are pulled out and burnt. Drenching with wet ceresan or benlate has been found to be effective in the control of the diseases. Geranium affected by tip rot during the monsoon period is controlled by spraying copper fungicides.

Root knot nematode *(Meloidogyne hapla)* poses a serious problem, resulting in 50 % loss of the yield. Application of Aldicarb 10G at the rate of 16 kg per hectare reduces the incidence of root-knot on geranium plants.

23.4. MINTS (*Mentha* spp)

Family: Labiatae

The genus *Mentha* consists of about 25 species, of which better known species of commerce are

1. Japanese mint (**M.** *arvensis* **L**. var. *piperascens* Holmes)
2. Pepper mint (M. *piperita* **L**)
3. Common or spear mint (**M.** *spicata* **L**.)
4. Scotch spear mint (**M.** *cardiaca Baker)* and
5. Bergamot mint (**M** *citrata* Ehrh.)

A brief cultivation practices of the above crops is dealt here.

23.4.1 JAPANESE MINT

Mentha arvensis and *M. arvensis* var. *piperascens* (Japanese mint) yield the oil of Mentha on steam distillation which is the raw material required for the manufacture of menthol. It is actually a white crystalline needle shaped substance obtained by a series of fractional chilling. World demand is more than 24,000 tonnes annually and Brazil is the leading country in its production. This was recently introduced to India and is now popularly grown in Jammu, Uttar Pradesh and Punjab to an area of 1.50 lakhs ha by nearly 1, 00, 000 farming community. We are able to meet 75-80 % of world demand of menthol. Menthol is used as a flavouring agent, anti-pruritic, and coolant and as a carminative. The spent grass is useful in making pulp for the manufacture of hard boards and papers.

Climate and Soil

Japanese mint prefers a cooler climate than a hot tropical or semi tropical climate. During winter it remains dormant. It requires a climate with adequate and regular rainfall during the period of its growth and good sunshine during its harvesting.

Mentha arvensis prefers a deep, fertile, loose, moist soil which will allow free growth of underground runner. It can be grown in neutral or slightly alkaline soils having pH of about 8 but its preference is within a range of 6 to 7.

Varieties

The CIMAP, Lucknow has released four high yielding menthol rich varieties. **Himalaya (MAS 1)**- a selection from Thai bud sprout, contain 0.8 to 1.0 % oil. The oil has a menthol content of 81 % and a low congealing point, it has higher regeneration capacity in the second and subsequent harvest and delayed harvesting time does not affect oil production. Highly resistant to rust, leaf spot, powdery mildew disease. Other varieties are **CIMAP/ Hybrid-77 (Kalka)** can produce up to 350 kg of essential oil per hectare per year and the oil contains about 81.5 % menthol. Another mutant viz **RRL118/3** possesses higher herbage and oil yield coupled with higher content of menthol (80-90 %) has been released for cultivation. Another variety **Shivalik** is compact bushy growth, thick leathery leaves, high oil and menthol content. **Koshi (Sel.3)** has high leaf density, synchronous growth habit and resistant to diseases like leaf spot, rust and powdery mildew while **Gomati** is an early variety.

Preparation of Field

The field should have a good tilth which can be obtained by thorough ploughing. At the time of final ploughing, endrin dust (5 %) must be applied @ 50 kg per hectare to safeguard the crop against any soil borne pests.

Planting: Mentha is propagated by suckers. 500 kg of suckers are required to plant one hectare. Suckers are usually obtained from the field that has been planted during the preceding year. The suckers once removed should be kept moist under shade until it is planted in the main field. It is better that suckers are planted at the earliest for ensuring good survival.

The suckers should be set in furrows at 5 to 7 cm deep with rows spaced at 60-75 cm apart. The suckers should be cut into pieces of 10 to 12 cm length before planting and are planted at a distance of 15 cm in the furrows. About 500 to 600 kg of suckers are required per hectare.

Irrigation: Mentha crop requires considerable moisture distributed throughout the growing season. Being a shallow rooted plant, it is better to irrigate at frequent intervals. During summer, it may be necessary to irrigate the fields at weekly intervals.

Weeding: Weeding is one of the important inter cultural operations since yield of oil depends largely on the extent of its freeness from weeds.

Manures and fertilizers: Farm yard manure at the rate of 20-25 tonnes per hectare is recommended and this organic manure gives good response.

At the time of planting (as basal dose) 50 kg of N, 75 kg of P and 37 kg of potash is applied. Besides, 75 kg of nitrogen is applied in three equal split doses, first top dressing is when the plants are about 15 cm high and the remaining doses of nitrogen should be applied after each harvest.

Harvesting and Distillation

First harvesting commences about 120 days from planting. This coincides with the stage when the plants will be nearly in the full bloom stage. Care should be taken that the crop is harvested in sunshine at a height of 4 to 8 cm from the ground level. Rainy days during harvesting period result in reduced oil yield. Therefore, it is better to avoid maturity time to coincide with the monsoon period. Planting of suckers is hence recommended by the end of February under North Indian Plains so that first cutting is over by middle of June, second cutting is taken from the beginning of September (before the monsoon season) and third cutting is also possible by November. The crop there after enters dormancy during winter months.

Steam distillation is good for extraction of mentha oil. The herbage should be allowed to wither for 12-24 hours before distillation. On an average, 25-30 kg of oil can be obtained in the first year and 20-25 kg in the second year.

Plant Protection

Termite attack is normally seen during dry months. The affected plants have yellow leaves and die subsequently. To control this, flood watering for 36 hours followed by soil application of 3 % haptafan @ 50 kg/ha as prophylactic measure is recommended. Besides, hairy cater pillar attacks this crop which can be controlled by dusting with 10 % Malathion.

22.4.2 PEPPER MINT

Pepper mint (*Mentha piperita* L.) is a perennial, glabrous, strongly scented essential oil yielding plant. It is a native of Mediterranean countries. It is cultivated in the temperate regions of Europe, Asia, North America and Australia. In India, there is a scope to cultivate this crop in the hills of temperate and warm temperate zones like Nilgris and Upper Pulney hills of Tamil Nadu besides in the plains of North India. The oil is used in the pharmaceutical and flavouring industries.

Climate and Soil

Pepper mint requires temperate to subtropical climate. It prefers long day conditions and requires an annual rainfall of about 100-120 cm. It can withstand frost also. It grows on a wide range of soils, but thrives best in deep well drained soils rich in organic matter. When compared to heavier soils, sandy or loamy soils produce oil of high quality. Neutral to slightly acidic soil is considered good.

Varieties

CIMAP, Lucknow has developed many improved cultivars as indicated below:

Varieties	Characteristics
Kukrail	High herb and high oil yield with superior oil quality
Pranjal	Higher, tolerance to Bihar hairy caterpillar. Oil content – 0.55 %, Herbage yield 0.31 q/25 m^2
CIM Madhuras	Sweet smelling chemo type, Menthol 31.23 %, Methyl acetate – 4.12 %, Menthone – 24.26 %, Oil content 0.65 %, Herbage yield 0.51 q/25 m^2
CIM Indus	High herb and high oil yield with superior oil quality, *Menthofeuron* type.
Tushar	Oil content 0.63 %, Herbage yield 0.47 q/25 m^2

Planting

The crop is propagated by underground parts called suckers. These suckers are collected from the previous crop and planted at a spacing of 45 × 30 cm in rows. Though planting can be done from the end of December to March, second week of February is considered best for higher herbage yield. Delay in planting results in reduction of herbage and oil yield.

Manures and manuring: Growing of a green manure crop prior to mint seems to be good. The crop responds well to application of Nitrogenous fertilizers and slightly to phosphatic and potash fertilizers. Application of 120 kg N, 60 kg each of P and K are considered optimum per hectare per year. P and K are applied as basal and N alone is applied in three equal split doses viz. 1/3 at planting, 1/3 60 days after planting and the remaining 1/3 after first cutting.

Weed control: Weed control is an important operation for the success of the crop. Due to slow sprouting and slow growth rate of the crop in the initial stage, weed seeds germinate and establish long before the crop comes. As manual weeding is not very effective, chemical weed control now is recommended. Pre-emergence application of terbacil @ 1.5 kg a.i/ha and application of propanil @ 1.75 kg a.i/ha effectively control the weeds in the field of *M. piperita*.

Irrigation: Irrigation immediately after planting and harvesting is essential. In the plains of North India, the crop requires to be irrigated at fortnightly interval frcm March to June, however, a week prior to harvest, it has to be stopped.

Harvesting and Distillation

The right stage for harvest is very important to distill quality oil. Harvesting is generally done when the mint is in bloom stage to get optimum oil yield and menthol content as the major percentage of essential oil formation is during the reproductive stage than vegetative or post reproductive stages. Under Indian conditions, optimum time for first cutting is between the middle of June to middle of July, (149 to 170 days from planting). The second and third cuttings are taken 65 to 85 days after first and second cuttings respectively.

The cut herbage is allowed to dry slightly. Excessive drying must be avoided to prevent loss of oil. Steam distillation is followed. Care must be taken to properly distill the oil as improper distillation gives a strongly burnt odour to the oil. The recovery percent generally varies from 0.10 to 0.20 percent on fresh weight basis.

Yield

The herbage yield is high during the first year and thereafter yields obtained in the second and third year are not economical due to so many factors. Hence, it is recommended to grow as an annual crop with a crop rotation of potato or vegetable etc. On an average, 10 to 12 tonnes of herbage yield per hectare can be obtained in the first year which on distillaiton may yield 55 to 80 kg of oil.

The oil quality is judged by many of its constituents, the chief being menthol (25 to 40 %), menthone (30 to 40 %) and methyl acetate (2.5 to 6 %).

23.4.3 SPEAR MINT (*M. spicata*)

The spear mint (*M.spicata*) formerly called as *M.viridis* is having carvone (70 to 80 percent) as its chief constituents. It is grown in about 1200 to 1500 ha, annually producing 80 to 100 tonnes of spearmint oil As the production is higher than the requirement, it is exported now. CIMAP has released two improved strains viz. **MSS 1** and **MSS 5** which is high yielding in terms of herbage as well as oil yield. Y.S.Parmar University of Horticulture and Forestry, Nauni, Solan has developed **Punjab Spearmint-1** with a herbage yield of 96.0 q/ha and higher oil content (= ~ 0.50 %) which is suitable for sub-temperate and sub tropical region. CIMAP, Lucknow has released two improved cvs **Arka** and **Neera** which are high yielding and superior quality oil. Recently an inter specific hybrid between *M. arvensis* and *M. spicata* (**Neer Kalka**) has been developed.

The cultural requirements of this crop are almost similar to *M. piperita.* It produces on an average 20 tonnes of herbage which on distillation yields 17 Kg of oil. The oil recovery varies from 0.11 to 0.14 per cent. The oil is widely used in pharmaceutical industry and also as flavouring and sweetening agent.

Another species viz *M.cardiaca* called scotch spearmint is also a good source of spearmint oil and it contains 65.00 per cent of carvone.

23.4.4 BERGAMOT MINT

Mentha citrata Ehrh commonly known as bergamot mint yields an aromatic oil on distillation which is rich in linalool and linalyl acetate. It is native of Europe and it is in cultivation in recent years in sub temperate and sub-tropical regions of our country. India requires about 40 tonnes of *M. citrata* oil while our production is nearly 1/10 of the demand. The oil is one of the classic perfume materials. It blends well in almost any perfume composition. The oil is widely used in lotions, cream, powders and soaps.

Climate and Soil

It is a crop of temperate region but can be commercially grown in sub-temperate and sub tropical regions. It can also tolerate frost. It prefers loamy sandy soils or peaty soils rich in organic matter. Drainage is very important and hence clayey soils should be avoided. It can be grown in neutral to alkaline soils.

Varieties

CIMAP, Lucknow has developed a new variety, '**Kiran**' by mutation breeding which is tall and vigorous in its growth and it produces about 239 kg oil /ha with 48 % linalool and 37 % linalyl acetate

Preparation

In the plains of North India, the field is completely ploughed free of clods and ridges and furrows are formed. Under South Indian hills, the field is thoroughly dug out for planting. The crop can be propagated vegetatively by runners, suckers and stolons. Runners are the best propagating material under South Indian hill conditions. Stem cutting during rainy months can be also used to multiply as the planting material and in that case, each cutting should have alteast one leaf for successful rooting.

Optimum time of planting in the plains of North is January while in the hills it is February - March and in South India hills, it is during May-June. A spacing of 40 × 40 cm or 40 × 30 cm is adopted.

Manures and fertilizers: It responds to application of nitrogen and phosphorus for herbage production but potassium has no influence on it. It has been found that NPK has no effect on the oil content of the plant. Nitrogen upto 120 kg and phosphorus up to 60 kg are recommended per hectare. Nitrogen is applied in three equal split doses, 1/3 at the time of planting, 1/3 60 days after planting and remaining 1/3 after first cutting.

Irrigation: The crop requires irrigation during summer or immediately after each cutting. As the peak growing season coincides with peak summer in the plains of North India, it is to be irrigated at fortnightly interval during such periods. It is also desirable to stop irrigation a weak prior to harvesting. It is grown as a rainfed crop in South Indian hills.

Weed control: Proper control of weed is essential for the successful growth of bergamot mint. Manual weeding atleast three times is practiced to keep the weeds under control. Recently, number of weedicides has been reported to be successful in checking the weed growth. Granular application of lasso @ 35 kg/ha as a pre-emergence weedicide is recommended specifically for bergamot mint.

Harvesting

M. citrata offers two to three cuttings per year depending upon the location, climate and other factors. In most of the places, the first cutting is taken during May to June as the delay results in excessive leaf shedding, second harvesting is taken during August/September and third harvest during October/November. To boost up the growth, application of GA at 200 ppm is recommended. The crop should be distilled in the fresh state as early as possible as delay in distillation affects the quality of oil. The oil recovery ranges from 0.20 to 0.60 % depending upon many factors such as place of growing, season of harvest etc. Oil yield per hactare varies from 75 to 150 kg per year. The oil contains about 40 to 60 percent of linalool and 9 to 27 percent of linalyacetate.

23.5. OCIMUM SP.

Family: **Labiatae**

The genus *Ocimum* is one of the important aromatic groups of herbaceous plant belonging to the family Labiatae. Many species of *Ocimum* contain various economically important essential oils used in perfumery and cosmetics industries. The major constituents in Ocimum oils include linalool, geraniol, citral, camphor, eugenol, methyl chavicol, safrol, thymol, methyl cinnamate etc. Ocimum species are used as herbs and find diverse uses in the indigenous systems of medicine in countries like India, Africa, Arabia, Australia, Malaya, Pacific Islands and Sri Lanka. The oil of certain species of *Ocimum* has the antifungal, bactericidal and insec-ticidal properties too.

Besides from the industrial point of view, Ocimum species with oil rich in camphor, cirtral, geraniol, linalool, linalyl acetate, methyl cinnamate, eugenol, thymol etc. can be harnessed to yield new commercial products. Ocimum species like *O. kilimandscharicum* and *O. canum,* being rich in natural source of camphor, attracted attention especially during second world war when there was an acute shortage of camphor. The products of Ocimum worth of 40 lakhs are annually imported into our country.

The genus *Ocimum* contains 160 species and is distributed in warmer parts of the hemispheres from sea level upto 1800 m. The maximum number of species is from the tropical rain forests of Africa and the centre of diversity in this genus are Africa, South America and Asia. The genus can be broadly divided into two groups viz. Basilicum group and Sanctum group based on the following differences.

	Particulars	Basilicum groups	Sanctum group
1.	Habit	Species are mostly herbaceous perennials	Biennials or triennials or woody perennial under shrubs
2.	Flower character	Bracts are petiolate flowers more conspicuous	Bracts are sessile flowers are small.
3.	Seeds	Mucilaginous	Non-mucilaginous

Floral characters of all *Ocimum* species are similar in basic structure. Flowers are protandrous. The duration between opening of the flower and dehiscence of anther and maturity of style varies between species to species. Flowers are mostly entomophilous, thus they are mostly out crossing types.

In India, *O. americanum*, *O. basilicum*, *O. canum*, *O. gratissimum* and *O. sanctum* are distributed. *O. gratissimum* is mostly cultivated in North India and *O. sanctum*, being considered as sacred by Hindus, is cultivated in almost every part of India.

The important species of *Ocimum* which can be exploited for their essential oil content are presented below:

Name of Species (1)	Major chemical constituents (%) (2)	Percentage of essential oil on F.W. basis. (3)
Ocimum canum (2n = 24)	Linalool (80-90)	2.0 to 4.0
O. canum (2n = 26)	Camphor (60-80)	2.5 to 5.0
O.basilicum var. *minima*	Geraniol (40-50) Eugenol (20-30)	0.8 to 1.5
O.basilicum var. *crispa*	Methyl chavicol (40-50) Linalool (25-30)	0.5 to 2.0
O. basilicum var. *glabratum*	Methyl chavicol (40-50) Linalool (20-25)	0.9 to 2.5
O. americanum	Methyl chavicol (70-75)	3.6 to 6.5
O. citriodorum	Citral (65-70)	1.5 to 2.5
O. kilimandscharicum	Camphor (70-80)	1.5 to 3.5
O. sanctum	Eugenol (60-75)	0.5 to 1.0
O. gratissimum	Eugenol (70-80)	0.5 to 1.5
O. viride	Thymol (60-70).	1.8 to 2.5

Varieties

The Regional Research Laboratory, Jammu has done extensive crop improvement programme in Ocimum and as a result evolved newer promising types which are now recommended for commercial cultivation. The important characters of these improved strains are presented below:

	Name of Species	Fresh herb Yield (t/ha).	Oil yield kg/ha	Oil recovery (Fresh basis) (%)	Major oil constituents
1.	*O.canum* RRL-01	40	200	0.5	Linalool (70-80 %)
2.	*O.america -num* RRL-02	41	185	0.45	Methyl chavicol (70-75 %)
3.	*O.viride* RRL-08	42	200	0.50	Thymol (65-70 %)
4.	*O.gratissi -mum* RRL-08	40	160	0.40	Eugenol (75-80 %)
5.	*O.basilicum* (RRL-07)	40	200	0.50	Citral (75-80 %)
6.	*O.basilicum* RRL-011	50	220	0.50 %	Linalool (40 v) and Methyl chavicol (35 %)
7.	Synthesized Amphidipl oid of Ocimum RRL-015	60	300	0.59 %	Eugenol (55-65 %)

Besides, CIMAP , Lucknow has developed the following varieties:

Varieties	Characteristics
(*Ocimum basilicum*)	
CIM - Soumya	Fresh herb yield 290 q/ha, oil content 0.68 % - fresh herb, 0.99 % from dry herb oil yield 197 kg/h Methyl chavicol 62.54%, Linalool -24.61 %.
Vikarsudha	A hybrid between exotic basil from Australia (Ec 331886) and a landrace (Badaun). Tall variety (75-90 cm). It is a photo insensitive and can be cultivated throughout the year except severe winter. The oil content is 0.7 %.
Kusmohak	Introduction from Argentina. Medium tall herb, high oil yield, superior oil quality. Higher linalool content (46 %) and moderate chavicol (38 %).

MC – 05	Biomass yield 36.4 t/h, Oil content 0.35 %, Oil yield 127.4 kg/ha , Methyl chavicol 87 %.
Sacred basil (*Ocimum sanctum*)	
CIM – Ayu	Plants of this variety are tall with green leaves.
CIM Angana	Plants of this variety are also tall but leaves are purple in colour.
CIM Kanchan	Methyl eugenol > 70 %, oil yield 94 kg/ha/year, Biomass yield 19.7 t/ha/year.

Recently an improved type of *O.gratissimum,* developed at Indian Institute of Integrated Medicine, Jammu as a substitute to eugenol yielding tree crops was named as **clocimum**, since the oil extracted from this *Ocimum* plant is clove scented. This was developed by hybridization between two chemo types of *O.gratissimum* differing in geographical distribution viz. Jammu type and African type and followed by repeated selection and intermating. This clocimum has a yield potential of 35 and 65 tonnes of fresh herbage in the first year and second year respectively. The oil recovery is about 0.5% on fresh weight basis, with a high eugenol content of 72 %.

Climate and Soil

As it is a tropical to subtropical plant, it prefers fairly to high rainfall areas with high humid conditions. Long day and high temperature condition enhance growth and higher oil production. Partial shade is not beneficial as it affects the oil content. It can be grown in hilly place up to 2000 m and it can tolerate drought as well as frost.

Ocimum can be grown in rich loam to poor lateritic soils. It can also come up in saline or alkaline soils to moderately acidic soils. Proper drainage is however essential as water- logging condition results in poor growth.

Field Preparation

The land is well prepared with two or three ploughing until a fine tilth of soil is obtained. Farm yard manure may be applied before the 2nd or 3rd ploughing. The crop is raised from seeds and can be grown annually from the middle of February to the end of September by direct sowing or transplanting.

1. **Direct sowing:** Seeds (200-250 g per ha) are mixed with sand to ensure an even distribution. Before sowing, the field is marked into rows 50 to 60 cm apart. Seed is sown by hand or drilled and covered with soil by disturbing the rows with foot taking care that seed remains at a depth of about 2 cm. in the soil. Seeds sown deeper fail to germinate. The field is irrigated after 24 hours depending on soil moisture. Germination starts within 10 to 15 days. After 20-25 days when the seedlings are 10-15 cm tall, first weeding and thinning, if necessary, is done.

2. **Transplanting:** Seeds are sown in the nursery towards end of March in raised bed. Germination is profuse and is complete in about 10 days.

Seedlings with 4-6 leaf stage become ready in about 6-7 weeks for transplanting. Seedlings may be transplanted at a spacing of 60 × 60 cm. The fields are irrigated both prior to and after transplanting. Rate of establishment of seedlings is good with only few mortalities.

Manures and manuring: Generally, no manure is required, but an application of 20–25 kg of N and 10-15 kg of P per hectare after one month from planting gives good herbage and oil yield. Another dose of N and P are applied at the same rate, 10-15 days after every harvest: Micronutrients given as foliar spraying especially 50 ppm of Cu and 100 ppm of Mo increase the total oil content *in O. sancturn.*

Harvesting

From sowing to first harvesting, it takes about 90-100 days in the case of direct sown and 75-90 days in the transplanted crop. The plants are cut 20-25 cm above the ground level in the first year, 20-30 cm level in second year and 35-45 cm level in the third year. Close cutting below 20 cm will be injurious to the plant. Immediately after harvest, the field has to be irrigated. The second and third harvests are taken at 50-60 days interval. Harvesting is generally done on a bright sunny day for better quality of oil.

Yield

On an average, 25 to 30 tonnes of herbage yield can be obtained per hectare in the first year and in the subsequent years it may even go up. The oil yield depends upon the species grown and many other factors such as stage of harvest, season etc.

23.6. PATCHOULI (*Pogestemon patchouli* Hook)

Family: **Labiatae**

It is an aromatic herbaceous perennial plant, yielding patchouli oil. Our country is importing annually patchouli oil worth rupees 50 lakhs to meet the internal demand. Indonesia is the largest producer of this oil in the world, accounting for 90 % of the world production.

Patchouli oil is used as a 'base' material in perfumery industry as it has strong fixative properties. It is also used as a flavour ingredient in many food products. It also possesses many medicinal properties.

There are three main species in this genus. *Pogestemon cablin* (Benth) Syn. *P. patchouli* has more commercial value since the oil content of the leaves is high compared to the other species. It varies from 2-6 percent, but in the other species viz *P.heyneanus* and *P.hortensis,* it varies from 0.05 to 2 percent only.

P. patchouli grows to a height of about 1 m and never flowers in Indonesia. It has hairless stem and twigs are quadrangular and have opposite leaves. The leaves are of a pale green colour, heart-shaped, slightly lobed and with downy hairs (more abundant on the lower surface than on the upper) and obtuse lobes and apex. The leaves have glands which secrete the real oil.

Varieties

Johore, Singapore and Indonesian strains were introduced. Johore yield best quality of oil. Singapore and Indonesian strains have high herbage and oil yield but oil quality is inferior.

Climate and Soil

It thrives the best in hot and humid conditions, therefore, can be grown in coastal areas of South India besides hill stations up to an elevation of 1500 m.

Patchouli might grow on a variety of soils, but virgin forest soils are the best. A deep loamy soil, rich in humus and nutrients, with good drainage but high water holding capacity without impervious layer in the topsoil is the most desirable soil type for optimum oil production.

Land preparation

The field must be thoroughly prepared to remove all obstacles, and this work must be done before the beginning of the rainy season. In the hill stations of South India, due to the sloppy terrain nature of the land, complete cleaning of the land is not recommended because it will encourage soil erosion. Therefore, slash weeding and spot cleaning are done where planting has to be exactly done.

Propagation

The real patchouli plant *(Pogestemon cablin* Benth) never flowers and therefore it has to be propagated vegetatively i.e. by cuttings, normally 15-20 cm long. Unrooted cuttings can also be planted straightaway in which case, 2 to 3 cuttings have to be planted per hill during rainy season. They will require initial artificial shading for establishment.

Cuttings from the middle part of a stem give better sprouting than from either end of the stem. Experiments conducted in Tamil Nadu showed that herbaceous cuttings of 20-25 cm length prepared from healthy top shoots with 4 to 5 pairs of leaves are the best material for propagation. August is the best month for preparation of such cuttings to have better rooting and survival in the field. Cuttings root better in sand medium.

The rooted cuttings about three weeks old may either be planted directly in the field or further raised in plastic bags and transplanted into the field later on. It will be sufficient to plant one cutting per hill.

Planting distance: The ideal planting distance is dependent on soil and climatic conditions. In Malaya planting is done in rows at 30 ×60 to 90 cm. In Tamil Nadu, a closer spacing of 30 × 30 cm is found to be ideal. On low land and damp clay soil, patchouli should be planted on ridges at 50 x 100 cm but on hilly land contour planting at 50 × l00 cm is advisable.

Intercropping: As patchouli requires hot and humid climate, it can be also grown as intercrop in plantation crops like rubber, coconut and coffee.

When intercropped in rubber or other plantation crops, it should be noted that oil content in leaves grown in shade tends to become less than in leaves grown in sites exposed to the sun. Experiments conducted at Horticultural Research Station, Thadiyankudisai also made clear that oil content gets reduced with shade.

Patchouli should normally be renewed after two to three years because of strong lignification and too low leaf production.

Manuring: No systematic manure is generally applied, but it is evident that considerable increase in yield can be obtained with the aid of fertilizers in combination with suitable methods of cultivation. Experiments conducted in Tamil Nadu show that a fertilizer dose of N 60, P_2O_5 30 and K_2O

30 kg/ha to be optimum for getting higher herbage and oil yield. Foliar application of urea soultions also promotes growth and leaf production, besides increasing the oil content of the leaves.

Harvesting

All parts of the plant, i.e., roots, stem, branches, stalks and leaves contain essential oil, but the oil content of the leaves is by far the highest. The oil is also contained mostly in the top three leaves (Youngest), therefore, it is recommended that the leaves be harvested when the plant has five pairs of leaves, and only the leaves are distilled as the low oil content in the stem and stalks renders them unsuitable for distillation. Therefore, only the tops of the plants should be cut, using scissors or sharp knives. If the entire plant is cut, regrowth would take too long and the interval between the harvests would also be prolonged. The length of the cut tops varies from 25-45 cm, depending on the height of the bush. Stalks of 25 cm long and 1/3 cm thick with all leaves attached are considered good distillation material.

The subsequent harvests occur at three-to five month intervals; the leaves may be harvested for two to three years and thereafter the plant must be replanted. To permit harvesting cycles throughout the year, the planting should be done in several stages. For this purpose rooted cuttings raised in plastic bags are suitable planting material, as planting can be spread over a long period.

After harvesting, the leaves and stalks are usually dried in the sun. The leaves are spread in thin layers on concrete floors or bamboo racks. Proper drying is of great importance for the quality of both leaves and oil. During drying the leaves are regularly turned over by hand or by means of a stick to promote even and thorough drying and prevent fermentation. Depending on sunshine and the relative humidity of the air, drying of patchouli leaves requires about three days. Therefore, during the drying process it is most important to avoid fermentation, which readily takes place if the leaves are not properly spread and turned over frequently.

Distillation: Processed leaves alone should be used for distillation. It is advisable to interchange high and low steam pressure, thus giving full range to the forces and hydro-diffusion, which are important in the distillation of dried plant material. Oil recovery generally ranges from 3 to 3.5 per cent. The major constituent of oil is *patchouli* which varies from 30 to 40 % in patchouli oils.

Plant Protection

Patchouli is often attacked by *Pachyzancla stultalis,* a leaf-roller and leaf-eating grasshoppers (Orthoptera) and leaf-eating crickets (Gryllidae) also have been noticed, the latter especially with young plants. In general, these pests can be easily controlled by spraying with 0.5 per cent dieldrin 50 per cent WP.

A more serious problem is nematodes, especially *Meloidogyne incognita, M. javanica, M. hapla* and *Pratylenchus brachyurus.* The leaves turn yellow and the plant starts wilting; young plants may be killed by the attack.

Control measures for nematode consists of following proper crop rotation especially with a non host green manure crop like *Mucana puricata,* growing along with banana, grapes and other crops which normally receive nematicide applications. Application of aldicarb 1.0 Kg a.i and carbofuran at 1.0 Kg a.i/ha are relatively best to reduce the population of root knot nematode to half. Application neem cake (1.0 t/ha) also is effective in controlling the population of nematode.

23.7. EUCALYPTUS SP.

Family: **Myrtaceae**

Eucalyptus belongs to the family of one of the most important trees of world. They are native of Australia. The term **Eucalyptus oil** denotes three distinct group of essential oils viz medicinal-type, perfumery-type and phellandrene-rich type, Blue gum *(Eucalyptus globulus)* belongs to the first group which has high cineole content and wide medicinal application. *Eucalyptus citriodora* Hook belongs to the second and third group which is characterized by their high citronellal and phellandrene content respectively.

E. globulus is a native to Tasmania and now it is planted practically all over the world. In India, it was first introduced in 1843 in Nilgiris, now it is cultivated in the states of Tamil Nadu, Karnataka and Kerala. *Eucalyptus globulus* is also a good source of pulp wood and fuel.

E. citriodora is a large tree, often attaining a great height with a smooth whitish to pale pink bark. It is commonly called as *citron scented gum* or *lemon grass scented gum.* It can be easily identified by its characteristic fruit and lemon scented leaves. Two strains viz. pubescent and glabrous type exist in *E. citriodora* and the former type is rich in oil content and oil quality. It is native of Queensland and is now grown in most of the hill stations of South India. The oil is widely used in soap, perfumery and cosmetic industries and for the isolation of 'hydroxy citronellal' used in the manufacture of high grade perfumes. The demand for the oil in our country is around 55 tonnes per annum.

Climate and Soil

The plant is sensitive to severe frost and excessive drought. It tolerates rainfall up to 400 cm but can be grown in places receiving rainfall from 200 to 300 cm annually. *E. globus* can grow well above 2000 m MSL but *E. citriodora* grows well from 1500 to 2000 m MSL, higher the altitude, better is the quality of the oil. It can be grown in acidic soil, rich in organic matter, with good drainage.

Nursery

Eucalyptus is propagated by seed. As the root system is sensitive to transplanting, the seeds are directly raised in polythene bags of 22 cm × 16 cm size. The containers are filled with pulverised shola soils. Two seeds are sown in each bag and the right time of sowing is January/February under South Indian conditions. The polybags are staked in the nurseries and partial shade is provided. The seeds normally take 10-15 days for germination and they attain plantable size within 2 to 3 1/2 months from sowing or when the plants have produced 5 to 6 true leaves.

Planting

The land is cleared of the vegetation and pits 30 × 30 × 45 cm size are dug at spacing of 2m × 2m. The pits are then allowed to wither before planting. The pits are filled with top soil after adding 30 g of rock phosphate per pit. Right planting season is the commencement of South West monsoon and while planting, polythene bags are completely removed and planted without damaging the root system. Staking the plants to permit the wind damage is desirable. Gap filling is done until two years to ensure proper population in the field.

Manures and manuring: The Cinchona department in Tamil Nadu recommends 200 g per plant a fertilizer mixture of NPK at 25:38:38. This fertilizer mixture is applied by pricking the soil to a depth of 8 cm during the end of monsoon season.

Harvesting and Distillation

These differ widely between these two species. In *E. citriodoro,* harvesting consists of pruning and collecting the terminal branchlets and leaves up to three years. In the fourth year, coppicing the main stem is done 5 cm above the stem portion having lignin. The coppicing cycle is adopted for every fourth year and the leaves are collected for distillation. Further it has been reported that instead of harvesting at the end of coppicing cycle, harvesting the leaves at periodic intervals (6-12 months) results in higher leaf yield with higher citronellal content.

Another method of harvesting involves pollarding the main stem at a height of 3 m and the leaves are regularly collected for distillation from the shoots that emerge from the pruned stems. Best time for harvest of the leaves is March-May as the leaves have high oil content at that time. In Wynad area, harvesting twice i.e. premonsoon period (May) and post monsoon (November) is recommended. The harvested leaves are dried in shade for one day and distilled. Steam distillation is preferred to other types of distillation. On an average, the leaves yield 1.0 percent oil. The oil is a rich source of citronellal (70 to 80 percent).

In the case of *E. globulus,* leaves are collected from the trees by cutting the side twigs twice or thrice in a year or the fallen leaves are collected from the plantations for distillation. When the plantation is felled for pulp wood purpose also, available leaves will be collected for distillation. The leaves are distilled high throughout the year but the most favourable time for distillation is from April to September, because of the yield of oil and cineole content during this period.

The collected leaves are dried in shade for three days and then subjected to steam distillation. Distillation per charge takes 5 to 7 hours. The yield of essential oil ranges from 0.75 to 1.25 percent. As the crude oil is wet, coloured and contain lower aliphatic aldehydes of unwanted odours, it has to be rectified or purified. This rectification involves the treating the oil over anhydrous sodium sulphate and distilling over 1 to 2 % caustic soda. The cineole content varies from 75 to 85 per cent.

The rectified oil is colourless and has an aromatic camphoroceous odour.

23.8   LAVENDER (*Lavendula* sp.)

Family: **Labiatae**

Lavender, oil of commerce is obtained from two species of *Lavendula* viz. *L. angustifolia* syn *L. officinalis)* which is commonly called as *True Lavender* and *L. latifolia,* commonly called Spike Lavender. A hybrid of these two species is known as Lavindin or *L. hybrida.* The other important species which yield oil are *L. viridis* and *L. stoechas.*

France is the largest supplier of lavender oil in the world. It can be grown in Himalayan region and Nilgiris in India. India meets its entire requirements of Lavender oil from Europe by spending approximately Rs 25 lakhs per year on import of Lavender oil.

Oil of Lavender has stimulant and carminative properties and is used in hysteria, nervous headache and other nervous infections. It is chiefly given in the form of the official compound tincture and spirit of lavender. It is also commonly employed for scenting evaporation lotions, as well as ointments, and liniments. The more important consumption of oil of lavender, however, is in perfumery.

Lavender is a shrub of 30 to 90 cm high with a short but irregular, crooked, much branched stem, covered with yellowish grey bark which come off in flakes, and very numerous, erect, straight, broom-like, slender, bluntly opposite, entire, sessile and linear 3 cm. Flowers very shortly stalked, 3-5 together in little opposite cymes in the axils bracts.

Climate and Soil

Lavender prefers loose, calcareous soil. It also thrives in most luxuriantly rock crevices in its native home. They can be grown in peaty soils of hill station but requires good drainage.

Lavender generally prefers dry and cool climate. In the hill stations, higher the altitude better is the quality of the oil. Certain species like *L. stoechas* is xerophilous which prefers arid conditions.

Varieties

IIIM, Jammu has developed a superior clone viz. **CIMAP/B-15 (Sher-e-Kashmir)** which gives 100 per cent higher oil yield than the existing types and also possesses 20 percent more oil content with higher linalyl acetate content.

Planting and Cultivation

Lavender may be planted by any one of the following methods:

Seed: Since lavender seed possesses a very hard pericarp, it must be kept at room temperature in humid sand or sawdust, for some time to bring about germination. As soon as the young shoots emerge, they are transplanted into nurseries and watered frequently. After several months, the plants can be transplanted into the open field.

Cuttings: This method does not always meet with success very well. The cuttings should be raised in nurseries for rooting before transplanting to the fields. Maintenance of cuttings without wilting is very difficult and hence requires frequent irrigation.

Young wild plants: Naturally growing young plants are also used for propagation. Such young plants are then raised in the nursery for an about six months, frequently watered and transplanted to the open fields. Recently, micro propagation has been also attempted in this plant using shoot tips and nodes as explants.

Planting

Lavender was planted earlier at a distance of 1.25 to 1.50 m between the rows and 0.60 to 0.75 cm. between the plants; recently it was found that 1.70 to 2 m between the rows and 0.50 to 0.60 m between the plants give better results. This means approximately 10,000 plants per hectare. Closer spacing helps to check the weed growth. On the other hand, a wider distance between the rows is preferable for inter cultural operations. General cultivation during the first two years includes one annual ploughing followed by one or two superficial tilling with a cultivator.

No recommended manurial schedule is available for lavender, but it has been found that Lavender responds to nitrogen and phosphorous application.

Harvesting

Although the yield is insignificant in the first and less in the second year, the young plants must be cut, care being taken not to injure them. From the third year onwards, it produces a sizeable crop which increases during the fourth and fifth year. In Kashmir, during the first year of planting, the growth and flowering is very poor. However, the yield increases steadily from second and third year of planting. In a well-developed plantation, a skilled cutter can daily collect several hundred kilograms of plant material (flowering tops and stalks).

Distillation

The leaves have to be dried in shade first. Dried or semidried lavender can be distilled within 40-45 minutes. The oil yield by steam distillation is about 0.81%. The oil contains 50-53 percent of ester.

The yield of oil depends upon many factors particularly climate and weather, altitude, method of cuttings, the condition of the plant material (fresh/dry) and the method of distillation.

Plant material cut during dry and sunny days yield more oil. At the beginning and towards the end of the harvest, the yield is lower than during the period of full bloom. The altitude also has some influence on the yield.

The specific gravity of the oil ranges from 0.88 to 0.896. The oil chiefly contains linalool and linalyl acetate. Spike lavender oil and lavindin oil also contain linalool, cineole, and camphor and linalyl acetate.

Third year plantation yields on an average 2000 kg of herbage per hectare.

23.9. SANDAL WOOD (*Santalum album* L.)

Family: **Santalaceae**

Sandal wood is the source of world famous Indian sandal wood oil, which is extensively used in the perfumery industry. It is a small evergreen tree, often growing to a height of 15 to 18 m and 2 to 2.4 m in girth. A good portion of the oil produced in the country is exported to a number of foreign countries, particularly USA, UK and other European countries.

In India, the tree flourishes from sea level up to 1350 m but the formation of heart wood is best between 600 to 900 m, with an annual rainfall of 850 to 1200 mm. A comparatively cool climate, moderate rainfall and sunshine are the ideal conditions for the development of the heart wood.

Sandal tree starts flowering from 3rd or 4th year and flowering season generally lasts from February to April while fruiting takes place between July and October. As this tree produces more seed, they are dispersed freely by birds and germinate profusely immediately after monsoon under natural conditions. Sandal is an obligatory root parasite with all species with which it comes in contact. The sandal tree produces seeds in large numbers, which are dispersed freely by birds and which germinate to give seedlings in large numbers following monsoon. These seedlings try to regenerate if they are protected from natural mortality factors like excessive heat, drought, fire, browsing and trampling.

Artificial regeneration can be done by the following methods:

1. Dibbling the seeds in the bushes and pits
2. Sowing the seeds on mounds in the trench mound plantations
3. Raising the seedlings in nursery in polybags and their transplantation in the field.

In all the cases of artificial regeneration, the seeds of host plants such as *Cassia siamea, Calotropis gigantea, Cajanus cajan* are dibbled in the ratio 1:3.

Harvesting

In the sandal wood, sap wood and heart wood are well demarcated. The scented heart wood is the most valuable portion. As the root is also scented, harvesting consists of uprooting them. The heart wood formation is good in trees of 30 to 60 years attaining a girth of 40-60 cm. The yield of the heart wood varies from locality to locality and the age of tree. On an average, a tree with 50-60 cm girth may yield 19 to 50 kg of heart wood.

Distillation of Oil

Major portion of the sandal wood oil of commerce is produced by steam distillation of the pulverized heart wood and root. Distillation generally requires 48 to 72 hours. The yield of oil varies from 1.5 to 2 % in the chips which constitutes an admixture of heart wood and sap wood and as high as 10 % by weight in the roots. The main constituents of sandal wood oil are α and β santalols which account for 90 to 93 % of the oil.

24

Other Economic Aromatic Plants

S. No.	Common name	Botanical name & family	Brief description
1.	Ambrette (Kasturi bhendi)	*Abelmoschus moschatus* Malvaceae	An annual shrub; seed is the economic part, the oil obtained by steam distillation is a clear yellow to amber coloured liquid with a strong musky odour, mainly used in the alcoholic and non-alcoholic beverages.
2.	Mexican Lavender (Linaloe)	*Bursera delpechiana* Burseraceae	Shurb to tree in stature, native of tropical and subtropical America, this tree does not become oil bearing until it is atleast twenty years old, wood contains the oil, incisions is often made on the trunk and branches to stimulate the development of the oil.

S. No.	Common name	Botanical name & family	Brief description
3.	Hops	*Humalus lupulus* (L) Moraceae	It is a climber, cool season crop, female flowers atkins and the dried glandular trichomes are the economic part, which contain a number of chemical constituents like soft resins, essential oils (0.25%), tannins etc. which form an essential ingredient in the manufacture of beer throughout the world.
4.	Liquorice	*Glycyrrhiza glabra* (L) Leguminosae	The underground portion consisting of long roots and thin rhizomes or stolons is the economic part, the chief constituent is the saponin like glycoside, glycyrrhizin (5 to 7%), it is five times sweeter than sucrose, major use is as a flavouring agent, especially in tobacco industry and in pharmaceutical industry, and its extract is used as a demulcent and expectorant
5.	Kapur or Karpur or Karpuram	*Cinnamomum camphora* (L) Nees & Eberm Lauraceae	A large handsome evergreen tree, can be cultivated in the South Indian bills upto 2000 m. above sea level; trees of 2-40 years of age can be harvested, and it responds to coppicing, on an average whole tree contains 0.8% camphor and 1.5% camphor oil (oil contains 50% camphor in solution)
6.	Pandanus or Kewda	*Pandanas fascicularis* (Lam.) Pandanaceae	The plant occurs wild in the coastal belt of east coast, flowers are collected early morning and by water distillation methods, the oil is extracted and is collected in a vessel having sandal wood oil to absorb the aroma of the flowers. Kewda attar

S. No.	Common name	Botanical name & family	Brief description
			and Kewda water are the other products prepared. Kewda attar is used in tobacco preparations especially for high grade jarda. Kewda water is used for flavouring syrups, soft drinks and other food preparations.
7.	Deodar Devadaru	*Cedrus deodara* (Roxb) Loud. Pinaceae	A graceful, evergreen tree growing to a height of 50 m or even high, grows extensively on the slopes of Himalayas the essential oil (1.5 to 6%) is produced from saw dust, wood shaving and stumps by steam distillation for a duration of 10-12 hours, oil is used in perfumery, room spray and as disinfectant etc.
8.	Cyperus	*Cyperus rotundus* (L) Cyperaceae	A weedy species found growing wildly in marshy and wet land conditions. It bears a dark brown coloured rhizomes, which on distillation yield 0.30% oil, which is used as as fixative by Indian perfumers for flavouring of tobacco and scenting of soaps.
9.	Rosemary	*Rosemarinus officinalis* (L) Labiatae	A perennial temperate shrub, grows wildly in the Mediterranean, also cultivated extensively in France, essential oil is distilled from fresh leaves and twigs, oil recovery 1.0-1.5%, chief constituent being comphene (11.2%) and 1,8-Cineole (19.2%), used in scenting of soaps, shampoos, and other toiletries, also used in perfuming of insecticides, deodorants, room sprays, and also used in flavouring food

S. No.	Common name	Botanical name & family	Brief description
			products and in pharmaceuticals preparations. Ooty-1 Rosemary- is resistant to leaf blight disease and white flies and aphids. Leaf : 12.4t/ha/year. Essential oil content 0.9%.
10.	Artemisia	*Artemisia annua* (L) Asteraceae	A wild plant found in Europe, the essential oil yield ranges from 0.3 to 0.4%, its chief constituent being Artemisinin (25 to 40%), the oil is used in cosmetic and perfumery industries.
11.	Clary Sage	*Salvia sclarea* (L) Labiatae	A small perennial and aromatic herb, requires cool climate. The green parts of the plant especially the flowering top contains an essential oil of delightful odour, the oil content varies from 0.016 to 0.135% depending upon the stage of flowering, the chief constituent being linalool (34%) and linalyl acetate (38%), valued in perfumery, the flower tops can be also extracted with volatile solvents (petroleum ether) to extract concrete (0.005%) and then to absolute by further purification.
12.	Davanam	*Artemisia pallens* (Walls) Asteraceae	Annual herb, about 60 cm tall, native of India and found grown in southern India, leaves possess a characteristic bluish-green colour and very inconspicuous flowers, the essential oil is obtained by the distillation of the aerial partsof theflowering herb, the chief constituent of the oil is 'Davanone', the oil is an expensive one and hence

S. No.	Common name	Botanical name & family	Brief description
			found in expensive perfume compositions. The oil is now exported to the western countries.
13.	Pine	*Pinus longifolia* Roxb. Pinaecae	An evergreen tree occurring in Himalaya and also growing in South Indian hills above 2500 melevation, yields an oleoresin gum which on distillation yields an essential oil (15-20%) known as oil of turpentine and a non-volatile product called ' Rosin', the oil contains chiefly a 'pinine' and (5 'purine'). This oil is widely used as a solvent in paint, varnish, and boot polish industries, also employed in pharmaceuticals, perfumery and for making synthetic camphor and insecticides.

25

Floral Concrete and Other Aromatic Products

The flowers obtained from commercial flower crops like jasmines, rose and tuberose in addition to their conventional uses for garland making, hair adorning, veni making etc., are now being used for the extraction of the floral concrete or aromatic oil which are needed by the high cost perfume industry. The essential oil present in these flower crops can not be separated by simple steam distillation. Hence the solvent extraction method is practiced in which the odoriferous substances of the flower is allowed to be absorbed by a highly volatile chemical solvent like perfumery grade hexane, petroleum ether etc., and then the solvent is evaporated leaving behind the odoriferous principles, known as **concrete.** The floral concrete thus obtained contains some impurities like plant waxes, albumin, colouring matter etc. These are removed by further evaporation with absolute alcohol. The pure material thus obtained is known as **absolute.** Though several other methods of extraction such as maceration technique, expression method, enfluerage process etc., are known, the solvent extraction method is the most common one.

The solvent extraction method involves the following steps:

(a) Dissolving the perfume material in the solvent by treating the flowers with the solvent. The solvents normally used are Petroleum ether

342

or Food grade Hexane with boiling point from 60 to 80°C. The concrete extracted from the latter solvent is much preferred in Indian and foreign perfumery markets. In order to ensure the complete extraction of the natural perfume, the container having the flowers and solvent may be rotated slowly for about 5 minutes in the rotary type or extractor (Figure 23)

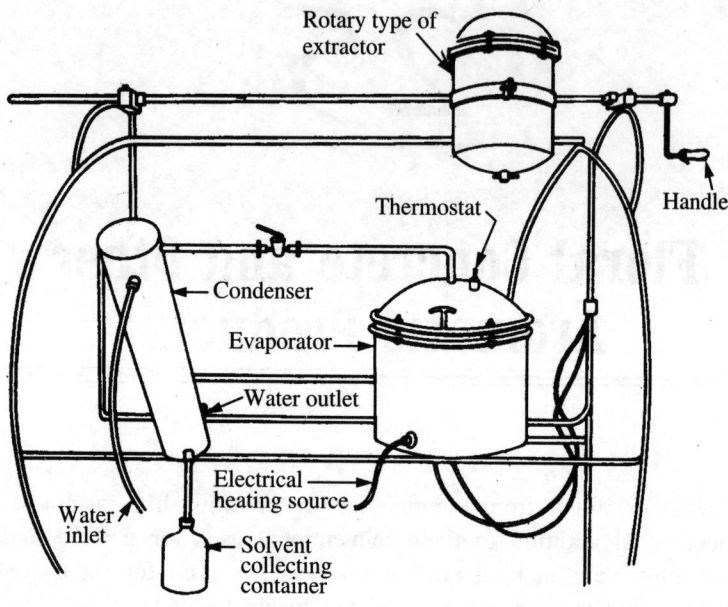

Figure 23 Pilot jasmine concrete extractor developed at TNAU, Coimbatore.

(b) The second step involves the removal of the solvent from the perfume material by evaporation. This is achieved by heating of the evaporator indirectly in a water bath till about 90% of the solvent is recovered by condensation. The remaining 10% of the solvent containing all the aromatic oil and waxy material is transferred to vacuum distillation unit where the complete removal of the solvent is effected and the concrete is left in the still in the form of molten wax. This concrete, when still hot is transferred to glass or aluminum containers for storage.

The floral concrete content of some of the important *Jasminum* sp. is as follows:

Common name	Botanical name	Concrete recovery (%)	Concrete yield (kg/ha)
Mullai	*Jasminum auriculatum* (Vahle)	0.28-0.35	13-28
Malligai	*J.sambac* (L) Ait.	0.14-0.18	1.1-15
Jathimalli	*J. grandiflorum*	0.25-0.32	13.5-20

Other Aromatic Products

Tuberose *(Polianthes tuberosa* L.), Edward rose *(Rosa borboniana)*, Red rose **(R. centifolia** L.) and Bulgarian Rose (or) Damask rose **(R. damascena (Mill.)** also yield concrete on solvent extraction but in India the follwoing products obtained from roses are commonly used:

1. Rose water - is obtained by the water distillation method. It is used as perfume for flavouring in confectionary, syrups and soft drinks. It is also used medicinally in eye lotions.

2. Rose attar - is obtained by water distillation of rose flowers and collecting the distillate over sandal wood oil. It is used as a perfume in agarbathies and also for flavouring of tobacco, particularly snuff and chewing tobacco.

3. Otto of rose (Rose oil) - is obtained by the water distillation of rose flowers by redistilling the distillate 2 or 3 times till it gets saturated with the oil dissolved in it. Then it is chilled and the oil drops floating on the surface of water are collected. The yield of oil comes to nearly 0.01 to -0.03% on fresh weight basis of flowers if improved type of distillation unit is used. It is one of the popular and valuable perfume materials and is extensively used in a number of perfumery formulations.

Annexure

Glossary of Some Medical Terms Used

Abortifacient	:	An agent that promotes abortion
Alterative	:	A drug which corrects disordered process of nutrition and restores the normal function of an organ or of the system
Anodyne	:	A drug that relieves pain
Anthelmintic	:	A drug that kills intestinal worms
Antipyretic	:	A drug which reduces fever
Antispasmodic	:	A drug which counteracts spasmodic disorders
Aphrodisiac	:	A drug which promotes sexual desire
Carminative	:	A drug which relieves flatulence
Demulcent	:	An agent having a soothing effect on the skin and mucous membranes
Diuretic	:	A drug which increases the secretion and discharge of urine
Emetic	:	A drug which induces vomiting

Expectorant	:	A drug that promotes the removal of catorrhal matter and phlegm from the bronchial tubes.
Fibrifuge	:	An agent used for reducing fever
Retrigerant	:	A drug which relieves feverishness or produces a feeling of coolness
Sedative	:	A drug which reduces excitement, irritation arid pain
Vermifuge	:	A drug which expels intestinal worms.

Index